The Contest Problem
Book VII

American Mathematics Competitions
1995–2000 Contests

© 2006 by
The Mathematical Association of America (Incorporated)
Library of Congress Catalog Card Number 2005937659
ISBN 0-88385-821-5
Printed in the United States of America
Current Printing (last digit):
10 9 8 7 6 5 4 3 2 1

The Contest Problem
Book VII

American Mathematics Competitions
1995–2000 Contests

Compiled and Augmented by
Harold B. Reiter

Published and distributed by
The Mathematical Association of America

MAA PROBLEM BOOKS SERIES

Problem Books is a series of the Mathematical Association of America consisting of collections of problems and solutions from annual mathematical competitions; compilations of problems (including unsolved problems) specific to particular branches of mathematics; books on the art and practice of problem solving, etc.

Council on Publications
Roger Nelsen, *Chair*

Roger Nelsen *Editor*
Irl C. Bivens   Richard A. Gibbs
Richard A. Gillman   Gerald Heuer
Elgin Johnston   Kiran Kedlaya
Loren C. Larson   Margaret M. Robinson
Mark Saul   Tatiana Shubin

The Contest Problem Book VII: American Mathematics Competitions, 1995–2000 Contests, compiled and augmented by Harold B. Reiter
A Friendly Mathematics Competition: 35 Years of Teamwork in Indiana, edited by Rick Gillman
The Inquisitive Problem Solver, Paul Vaderlind, Richard K. Guy, and Loren C. Larson
International Mathematical Olympiads 1986–1999, Marcin E. Kuczma
Mathematical Olympiads 1998–1999: Problems and Solutions From Around the World, edited by Titu Andreescu and Zuming Feng
Mathematical Olympiads 1999–2000: Problems and Solutions From Around the World, edited by Titu Andreescu and Zuming Feng
Mathematical Olympiads 2000–2001: Problems and Solutions From Around the World, edited by Titu Andreescu, Zuming Feng, and George Lee, Jr.
The William Lowell Putnam Mathematical Competition Problems and Solutions: 1938–1964, A. M. Gleason, R. E. Greenwood, L. M. Kelly
The William Lowell Putnam Mathematical Competition Problems and Solutions: 1965–1984, Gerald L. Alexanderson, Leonard F. Klosinski, and Loren C. Larson

The William Lowell Putnam Mathematical Competition 1985–2000: Problems, Solutions, and Commentary, Kiran S. Kedlaya, Bjorn Poonen, Ravi Vakil

USA and International Mathematical Olympiads 2000, edited by Titu Andreescu and Zuming Feng

USA and International Mathematical Olympiads 2001, edited by Titu Andreescu and Zuming Feng

USA and International Mathematical Olympiads 2002, edited by Titu Andreescu and Zuming Feng

USA and International Mathematical Olympiads 2003, edited by Titu Andreescu and Zuming Feng

USA and International Mathematical Olympiads 2004, edited by Titu Andreescu, Zuming Feng, and Po-Shen Loh

MAA Service Center
P. O. Box 91112
Washington, DC 20090-1112
1-800-331-1622   fax: 1-301-206-9789

Contents

Preface ix
46th AHSME, 1995 1
47th AHSME, 1996 9
48th AHSME, 1997 17
49th AHSME, 1998 25
50th AHSME, 1999 33
Sample AMC 10, 1999 39
51st AMC 12, 2000 45
1st AMC 10, 2000 53
50th Anniversary AHSME 59
46th AHSME solutions, 1995 71
47th AHSME solutions, 1996 83
48th AHSME solutions, 1997 95
49th AHSME solutions, 1998 107
50th AHSME solutions, 1999 119
Sample AMC 10 solutions, 1999 129
51st AMC 12 solutions, 2000 135
1st AMC 10 solutions, 2000 145
Additional Problems 153
Solutions to Additional Problems 159
Classification 175
About the Editor 183

Preface

History

Name and sponsors

The exam now known as the AMC 12 began in 1950 as the Annual High School Contest under the sponsorship of the Metropolitan (New York) Section of the Mathematical Association of America (MAA). It was offered only in New York state until 1952 when it became a national contest under the sponsorship of the MAA and the Society of Actuaries. By 1982, sponsorship had grown to include the national high school and two-year college honorary mathematics society Mu Alpha Theta, the National Council of Teachers of Mathematics (NCTM), and the Casualty Actuary Society. Today there are twelve sponsoring organizations, which, in addition to the above, include the American Statistical Association, the American Mathematical Association of Two-Year Colleges, the American Mathematical Society, the American Society of Pension Actuaries, the Consortium for Mathematics and its Applications, the national collegiate honorary mathematics society Pi Mu Epsilon, and the National Association of Mathematicians. During the years 1973–1982 the exam was called the Annual High School Mathematics Examination. The name American High School Mathematics Examination and the better known acronym AHSME, were introduced in 1983. At this time, the organizational unit became the American Mathematics Competitions (AMC), a subcommittee of the Mathematical Association of America. Also in 1983, a new exam, the American Invitational Math Exam (AIME), was introduced. Two years later, the AMC introduced the American Junior High School Mathematics Examination (AJHSME). In February 2000, the AMC introduced the AMC

10 for students in grade ten and below. At the same time, the AMC changed the name AJHSME to AMC 8 and AHSME to AMC 12. The two high school exams became 25 question, 75 minute exams.

Participation

Before 1992, the scoring of the exam was done locally, in some states by the teacher-managers themselves, and in other states by the volunteer state director. Beginning in 1992, all the scoring was done at the AMC office in Lincoln, Nebraska. Beginning in 1994, students were asked to indicate their sex on the answer form. The following table shows the degree of participation and average scores among females versus that for males.

Year	Females	Mean	Males	Mean	Unspecified	Mean
1994	104,471	68.8	120,058	76.0	6,530	70.6
1995	115,567	72.3	133,523	78.5	6,877	73.7
1996	124,491	65.8	142,750	71.2	6,659	67.8
1997	120,649	63.8	140,359	69.8	7,944	65.5
1998	108,386	66.5	128,172	71.9	7,438	67.8
1999	105,705	66.1	126,992	71.1	8,200	67.5
2000(12)	71,272	61.0	89,965	67.9	5671	64.3
2000(10)	49,288	60.8	52,836	67.5	4870	63.6

Related Exams

Until the introduction of the AIME in 1983, the AHSME was used for several purposes. It was introduced in order to promote interest in problem solving and mathematics among high school students. It was also used to select participants in the United States of America Mathematical Olympiad (USAMO), the six question, nine hour exam given each May to honor and reward the top high school problem solvers in America. The USAMO was used to pick the six-student United States Mathematical Olympiad team for the International Mathematical Olympiad competition held each July. With the introduction of the AIME, which was given the primary role of selecting USAMO participants, the AHSME question writing committee began to focus on the primary objective: providing students with an enjoyable problem-solving adventure. The AHSME became accessible to a much larger body of students. Some 7th and 8th graders, encouraged by their successes on the AJHSME, began participating.

Calculators

In 1994, calculators were allowed for the first time. At that time, the AMC established the policy that every problem had to have a solution without a calculator that was no harder than a calculator solution. In 1996, this rule was modified to read 'every problem can be solved without the aid of a calculator'. Of course the availability of the graphing calculator, and now calculators with computer algebra systems (CAS) capabilities has changed the types of questions that can be asked. Allowing the calculator has had the effect of limiting the use of certain computational problems. Referring to the Special Fiftieth Anniversary AHSME, problems [1954-38], [1961-5], [1969-29], [1974-20], [1976-30], [1980-18], [1981-24], and [1992-14] would all have to be eliminated, either because of the graphing calculator's "solve and graphing" capabilities or because of the symbolic algebra capabilities of some recent calculators. But the AMC has felt, like NCTM, that students must learn when not to use the calculator as well as when and how to use it. Thus questions which become more difficult when the calculator is used indiscriminately are becoming increasingly popular with the committee. For example, consider [1999-21] below: how many solutions does $\cos(\log x) = 0$ have on the interval $(0, 1)$? Students whose first inclination is to construct the graph of the function will be led to the answer 2 since in each viewing window, the function appears to have just two intercepts. However, the composite function has infinitely many x-intercepts.

Scoring

The number of problems and the scoring system has changed over the history of the exam. In the first years of the AHSME, there were a total of 50 questions with point values of 1, 2, or 3. In 1960, the number of questions was reduced from 50 to 40 and, in 1967, was further reduced from 40 to 35. The exam was reduced to 30 questions in 1974. In 1978, the scoring system was changed to the formula $30 + 4R - W$, where R is the number of correct answers and W is the number of wrong answers. In 1985, the scoring formula was $30 + 4R - W$, for a 30-problem contest. In 1986, the scoring formula changed to $5R + 2B$, where B is the number of blanks, again for a 30-question test. The formula and number of questions remained unchanged until the year 2000.

Beginning with the 2000 exams, some important changes took place. In order to accommodate school systems, the number of questions was reduced from 30 to 25 and the time allowed was reduced from 90 minutes to 75 minutes. The committee sought to continue to make the first five problems straightforward and the last five very challenging. The intension was to remove one question from what had been 6–10, one from what had been 11–15, and three from those in the range 16–25. The committee was also concerned that a bad experience with the AHSME might discourage capable younger students. The solution was a new exam, the AMC 10, specifically designed for students in grades 10 and below. The decision to include only topics that younger students might have seen was made. No trigonometry, logarithms, or complex numbers would appear on the AMC 10.

Also in the year 2000, the scoring formula changed to 6R + 2B for the 25-question test. In 2002, the formula changed again to 6R + 2.5B for a 25-question test. Students qualify for the AIME in much the same way as before with one exception. Because the exam was much harder in some years, the AMC decided to guarantee that at least 5% of the AMC 12 participants qualify for the AIME. The 'top 5%' rule was instituted in the year 2001. Since 2001, invitation to the AIME has been limited to students who 'scored 100 or placed in the top 5% among all scores on the AMC 12'. On the AMC 10 from 2000 to 2003, invitation was limited to the top 1%. In 2004 and 2005 the invitation to the AIME for AMC 10 participants was based upon a score of 120, or placement in the top 1% of scores on the AMC 10.

The 50th Anniversary of AMC

In 2001, three former AHSME chairpersons collaborated on an article, **The American High School Mathematics Examination: A 50 Year Retrospective**, which appeared in the journal *Mathematics Competitions*. Some of the material here is taken from that article. The article discusses some of the ways the AHSME changed over the years. For example, in the early years, there were often straightforward computational problems. See the 1950 problem on the Special Fiftieth Anniversary AHSME for a "rationalizing the numerator" problem on page 57. Many early problems involved the simplification of complex fractions, or complicated factoring.

Changes in problem types

It is interesting to see the how the test has changed over the years.

Preface

The table below shows how many problems of each of ten types appeared in each of the five decades of the exam and the percent of the problems during that decade which are classified of that type.

Classification of Problems by Decade

Classification	1950–59	1960–69	1970–79	1980–89	1990–99
All problems	500 (100%)	390 (100%)	320 (100%)	300 (100%)	300 (100%)
Geometry	203 (40.6%)	215 (55.1%)	178 (55.6%)	168 (56%)	100 (33.3%)
Logarithms	24 (4.8%)	18 (4.6%)	8 (2.5%)	10 (3.3%)	8 (2.7%)
Logic	7 (1.4%)	8 (2.1%)	4 (1.3%)	3 (1.0%)	6 (2.0%)
Combinatorics	7 (1.4%)	4 (1.0%)	7 (2.2%)	20 (6.7%)	32 (10.7%)
Probability	0 (0%)	0 (0%)	10 (3.1%)	20 (6.7%)	10 (3.3%)
Statistics	0 (0%)	0 (0%)	0 (0%)	14 (4.7%)	7 (2.3%)
Trigonometry	0 (0%)	0 (0%)	11 (3.5%)	17 (5.6%)	8 (2.7%)
Number Theory	14 (2.8%)	20 (5.1%)	41 (12.8%)	25 (8.3%)	21 (7.0%)
Absolute Value	4 (0.8%)	11 (2.8%)	24 (6.2%)	14 (4.7%)	5 (1.7%)
Function notation	6 (1.2%)	4 (1.0%)	10 (3.1%)	12 (4.0%)	13 (4.3%)
Function composition	4 (0.8%)	2 (0.5%)	4 (1.3%)	3 (1.0%)	5 (1.7%)

In the 1960s, counting problems began to appear. In the early 1970s, trigonometry and geometric probability problems were introduced. In the 1980s, more problems involving statistical ideas appeared: averages, modes, range, and best fit. Problems involving several areas of mathematics are much more common now, especially problems which shed light on the rich interplay between algebra and geometry, between algebra and number theory, and between geometry and combinatorics.

Some of the entries above need some elaboration. For example, a problem was considered a trigonometry problem if a trigonometric function was used in the statement of the problem. Many of the geometry problems have solutions, in some cases alternative solutions, which use trigonometric functions or identities, like the Law of Sines or the Law of Cosines. These problems are not counted as trig problems. A very small number of problems are counted twice in the table. Many problems overlap two or more areas. For example, a problem might ask how many of certain geometric configurations are there in the plane. The configurations might be most easily defined using absolute value, or floor, or ceiling notation (greatest and least integer functions). Such a problem could be counted in any of the three categories geometry, combinatorics, or absolute value, floor and ceiling. In cases like this, we looked closely at the solution to see if it was predominantly of one of the competing types. This situation often arises in the case of number theory-combinatorics problems because many of these types of objects that we want to count are defined by divisibility or digital properties encountered in number theory, but often invoke binomial coefficients to count. A few problems of this type are double counted. Many of the early problems are what we might call exercises. That is, they are problems whose solutions require only the skills we teach in the classroom and essentially no ingenuity. With the advent of the calculator in 1994, the trend from exercises (among the first ten) to easy but non-routine problems has become more pronounced. Note that even the hardest problems in the early years often required only algebraic and geometric skills. In contrast, many of the recent harder problems require some special insight. Compare, for example [1951–48], one of the three hardest that year, with number [1996–27]. The former requires a few applications of the Pythagorean Theorem, whereas the latter requires not only Pythagorean arithmetic, but spatial visualization and manipulation of inequalities as well.

Other Considerations

Not all types of mathematics problems are well-suited for either the AMC 12 or the AMC 10. Some problems require too much time. Also, some problems are simply not well-suited to the multiple choice format.

Students have an average of three minutes to work the problems. Therefore, problems that require considerable computation or experimentation cannot be used. Although the committee has developed considerable skill at crafting multiple choice problems, nevertheless some problems must be discarded. These include problems for which the key insight is the determination of an invariant in a process. Of course problems which ask the solver to provide a proof are not suitable. Thus, a section of **Additional Problems** is included in this book. Some of these problems appeared in the Sunday London Times brainteaser column and others appeared in the editor's problems column in *Mathematics and Informatics Quarterly*.

Acknowledgements

I wish to thank my wife Betty who supported and encouraged me through a difficult first year of the AHSME chairmanship when I was learning to use LaTeX and PicTeX. Betty steadfastly and carefully read each contest just before it was to be printed, often finding tiny errors that had been missed by the other volunteer reviewers.

I also want to thank my daughter Ashley Reiter Ahlin who motivated my interest in problem solving and in precollege mathematics in general. It is not a big exaggeration to say Ashley and I became problem solvers together during her elementary, middle, and high school years.

The chair of the American Mathematics Competitions during my tenure as AHSME problems chair was Richard Gibbs, for whose able and good-humored leadership I am very grateful. Dick is an ardent problem solver who was always there to discuss the problems. Leo Schneider, my predecessor as AHSME problems chair, deserves a great deal of credit for helping, not the least of which was the example he set for paying attention to details, providing LaTeX style files, and for providing a refined mechanism for producing the exam. During my six-year term as AHSME chair, I was fortunate to work with some very talented problem posers, both committee members and panelists. Without these terrific volunteer problemists, this book would not be possible. In fact, this volume is a compilation of the work of these committee members and advisory panel members who contributed the problems and then reviewed and polished them. These contributors include: Ashley Ahlin, Mangho Ahuja, Titu Andreescu, Joyce Becker, Don Bentley, George Berzsenyi, Janice Blasberg, Steve Blasberg, David Bock, Paul Bruckman, Bruce Brombacher, Tom

Butts, Gilbert Casterlow, Sheng Cheng, John Cocharo, Greg Crow, Jim DeFranza, MaryLou Derwent, Rad Dimitric, Fred Djang, David Doster, David Drennan, B. Ellington, Joseph Estephan, Daryl Ezzo, Doug Faires, Zumig Feng, M. Fogel, Susan Foreman, Zack Franco, Sister J. Furey, Gregory Galperin, Dick Gibbs, George Gilbert, Victor Gutenmacher, Bill Hadley, David Hankin, John Haverhals, Bryan Hearsey, Gary Hendren, Doug Hensley, Gerald Heuer, Dick Horwath, John Hoyt, Ruth Hubbard, Ann Hudson, John Jensen, Elgin Johnston, Dan Kennedy, Joe Kennedy, Clark Kimberling, Gerald Kraus, Sheila Krilov, Sylvia Lazarnick, John Lawlor, Ben Levy, Lewis Lum, Steve Maurer, Eugene McGovern, Walter Mientka, Robert Musen, Akira Negi, R. Newman, Anna Olecka, Kheim Ngo, Rick Parris, Pete Pedersen, K. Penev, Leonna Penner, Jim Propp, Stan Rabinowitz, J. Rauff, T. Reardon, Paul Reynolds, Steve Rodi, Franz Rothe, Cecil Rousseau, Richard Rusczyk, Mark Saul, Vince Schielack, Leo Schneider, Arlo Schurle, J. Scott, Steve Shaff, H. Shankar, Terry Shell, Charlyn Shepherd, Alice Snodgrass, Alex Soifer, Priscilla Spoon, P. Steinberg, Dave Tanner, M. Tent, Al Tinsley, Andre Toom, Radu Toma, Tom Tucker, Clair Tuckman, Gary Walker, Xiaodi Wang, Eric Wepsic, Susan Schwartz Wildstrom, Ronald Yannone, Peter Yff, and Paul Zeitz.

After the first two exams, I began to use Richard Parris's free Peanut software, which automatically produces PiCTeX output. Beverly Ruedi has been a great help with technical LaTeX problems and general questions about style. I am also grateful to Kiran Kedlaya and Gerald Heuer, whose careful reading of the manuscript uncovered several errors. I am especially grateful to Elgin Johnston, who pointed out several errors and also provided the Inclusion/Exclusion finishing touch on the problem Unsquare Party.

Also, I appreciate my UNC Charlotte and Davidson College colleagues who make problem solving a pleasant group activity. Here at UNC Charlotte, I have enjoyed such sessions over the years with Charles Burnap, T. G. Lucas, Gabor Hetyei and Stas Molchanov. At Davidson, Ben Klein, Irl Bivens and Stephen Davis are always eager to take on a new mathematical challenge.

Finally, I appreciate my friend and coauthor Arthur Holshouser, with whom I have worked for several years. Arthur is the most tenacious problem solver I have ever known. Arthur has generously given much of his time to some only mildly interesting problems.

46th AHSME, 1995

1. Kim earned scores of 87, 83 and 88 on her first three mathematics examinations. If Kim receives a score of 90 on the fourth exam, then her average will

    (A) remain the same (B) increase by 1
    (C) increase by 2   (D) increase by 3
    (E) increase by 4

2. If $\sqrt{2 + \sqrt{x}} = 3$, then $x =$

    (A) 1 (B) $\sqrt{7}$ (C) 7 (D) 49 (E) 121

3. The total in-store price for an appliance is $99.99. A television commercial advertises the same product for three easy payments of $29.98 and a one-time shipping and handling charge of $9.98. How much is saved by buying the appliance from the television advertiser?

    (A) 6 cents (B) 7 cents (C) 8 cents
    (D) 9 cents (E) 10 cents

4. If M is 30% of Q, Q is 20% of P, and N is 50% of P, then $M/N =$

    (A) $\dfrac{3}{250}$ (B) $\dfrac{3}{25}$ (C) 1 (D) $\dfrac{6}{5}$ (E) $\dfrac{4}{3}$

5. A rectangular field is 300 feet wide and 400 feet long. Random sampling indicates that there are, on the average, three ants per square inch throughout the field. [12 inches = 1 foot.] Of the following, the number that most closely approximates the number of ants in the field is

   (A) 500 thousand   (B) 5 million   (C) 50 million
   (D) 500 million   (E) 5 billion

6. The figure shown can be folded into the shape of a cube. In the resulting cube, which of the lettered faces is opposite the face marked x?

   (A) A (B) B (C) C (D) D (E) E

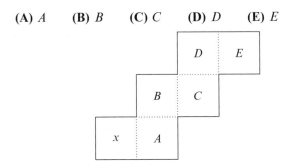

7. The radius of Earth at the equator is approximately 4000 miles. Suppose a jet flies once around Earth at a speed of 500 miles per hour relative to Earth. If the flight path is a negligible height above the equator, then, among the following choices, the best estimate of the number of hours of flight is

   (A) 8   (B) 25   (C) 50   (D) 75   (E) 100

8. In triangle ABC, $\angle C = 90°$, $AC = 6$ and $BC = 8$. Points D and E are on $\overline{AB}$ and $\overline{BC}$, respectively, and $\angle BED = 90°$. If $DE = 4$, then $BD =$

   (A) 5
   (B) 16/3
   (C) 20/3
   (D) 15/2
   (E) 8

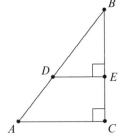

9. Consider the figure consisting of a square, its diagonals, and the segments joining the midpoints of opposite sides. The total number of triangles of any size in the figure is

   (A) 10
   (B) 12
   (C) 14
   (D) 16
   (E) 18

10. The area of the triangle bounded by the lines $y = x$, $y = -x$ and $y = 6$ is

    (A) 12 (B) $12\sqrt{2}$ (C) 24 (D) $24\sqrt{2}$ (E) 36

11. How many base 10 four-digit numbers, $N = \underline{a}\,\underline{b}\,\underline{c}\,\underline{d}$, satisfy all three of the following conditions?

    (i) $4,000 \le N < 6,000$; (ii) N is a multiple of 5; (iii) $3 \le b < c \le 6$.

    (A) 10 (B) 18 (C) 24 (D) 36 (E) 48

12. Let f be a linear function with the properties that $f(1) \le f(2)$, $f(3) \ge f(4)$, and $f(5) = 5$. Which of the following statements is true?

    (A) $f(0) < 0$ (B) $f(0) = 0$ (C) $f(1) < f(0) < f(-1)$
    (D) $f(0) = 5$ (E) $f(0) > 5$

13. The addition below is incorrect. The display can be made correct by changing one digit d, wherever it occurs, to another digit e. Find the sum of d and e.

$$\begin{array}{r} 7\,4\,2\,5\,8\,6 \\ +\,8\,2\,9\,4\,3\,0 \\ \hline 1\,2\,1\,2\,0\,1\,6 \end{array}$$

    (A) 4 (B) 6 (C) 8 (D) 10 (E) more than 10

14. If $f(x) = ax^4 - bx^2 + x + 5$ and $f(-3) = 2$, then $f(3) =$

    (A) -5 (B) -2 (C) 1 (D) 3 (E) 8

15. Five points on a circle are numbered 1, 2, 3, 4, and 5 in clockwise order. A bug jumps in a clockwise direction from one point to another around the circle; if it is on an odd-numbered point, it moves one point, and if it is on an even-numbered point, it moves two points. If the bug begins on point 5, after 1995 jumps it will be on point

(A) 1   (B) 2
(C) 3   (D) 4
(E) 5

16. Anita attends a baseball game in Atlanta and estimates that there are 50,000 fans in attendance. Bob attends a baseball game in Boston and estimates that there are 60,000 in attendance. A league official who knows the actual numbers attending the two games notes that:

   i. The actual attendance in Atlanta is within 10% of Anita's estimate.

   ii. Bob's estimate is within 10% of the actual attendance in Boston.

   To the nearest 1,000, the largest possible difference between the numbers attending the two games is

   (A) 10,000   (B) 11,000   (C) 20,000
   (D) 21,000   (E) 22,000

17. Given regular pentagon $ABCDE$, a circle can be drawn that is tangent to $\overline{DC}$ at D and to $\overline{AB}$ at A. The number of degrees in minor arc AD is

   (A) 72   (B) 108
   (C) 120 (D) 135
   (E) 144

18. Two rays with common endpoint O form a 30° angle. Point A lies on one ray, point B on the other ray, and $AB = 1$. The maximum possible length of OB is

   (A) 1   (B) $\dfrac{1+\sqrt{3}}{\sqrt{2}}$ (C) $\sqrt{3}$ (D) 2   (E) $\dfrac{4}{\sqrt{3}}$

19. Equilateral triangle DEF is inscribed in equilateral triangle ABC as shown with $\overline{DE} \perp \overline{BC}$. The ratio of the area of $\triangle DEF$ to the area of $\triangle ABC$ is

(A) $1/6$
(B) $1/4$
(C) $1/3$
(D) $2/5$
(E) $1/2$

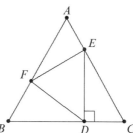

20. If a, b and c are three (not necessarily different) numbers chosen randomly and with replacement from the set $\{1, 2, 3, 4, 5\}$, the probability that $ab + c$ is even is

(A) $\dfrac{2}{5}$ (B) $\dfrac{59}{125}$ (C) $\dfrac{1}{2}$ (D) $\dfrac{64}{125}$ (E) $\dfrac{3}{5}$

21. Two nonadjacent vertices of a rectangle are $(4, 3)$ and $(-4, -3)$, and the coordinates of the other two vertices are integers. The number of such rectangles is

(A) 1 (B) 2 (C) 3 (D) 4 (E) 5

22. A pentagon is formed by cutting a triangular corner from a rectangular piece of paper. The five sides of the pentagon have lengths 13, 19, 20, 25 and 31, although this is not necessarily their order around the pentagon. The area of the pentagon is

(A) 459 (B) 600 (C) 680 (D) 720 (E) 745

23. The sides of a triangle have lengths 11, 15, and k, where k is an integer. For how many values of k is the triangle obtuse?

(A) 5 (B) 7 (C) 12 (D) 13 (E) 14

24. There exist positive integers A, B, and C, with no common factor greater than 1, such that
$$A \log_{200} 5 + B \log_{200} 2 = C.$$
What is $A + B + C$?

(A) 6 (B) 7 (C) 8 (D) 9 (E) 10

25. A list of five positive integers has mean 12 and range 18. The mode and median are both 8. How many different values are possible for the second largest element of the list?

   (A) 4   (B) 6   (C) 8   (D) 10   (E) 12

26. In the figure, $\overline{AB}$ and $\overline{CD}$ are diameters of the circle with center O, $\overline{AB} \perp \overline{CD}$, and chord $\overline{DF}$ intersects $\overline{AB}$ at E. If $DE = 6$ and $EF = 2$, then the area of the circle is

   (A) 23π
   (B) $47\pi/2$
   (C) 24π
   (D) $49\pi/2$
   (E) 25π

27. Consider the triangular array of numbers with $0, 1, 2, 3, \ldots$ along the sides and interior numbers obtained by adding the two adjacent numbers in the previous row. Rows 1 through 6 are shown.

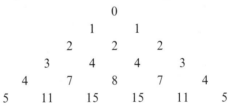

Let $f(n)$ denote the sum of the numbers in row n. What is the remainder when $f(100)$ is divided by 100?

   (A) 12   (B) 30   (C) 50   (D) 62   (E) 74

28. Two parallel chords in a circle have lengths 10 and 14, and the distance between them is 6. The chord parallel to these chords and midway between them is of length $\sqrt{a}$ where a is

   (A) 144
   (B) 156
   (C) 168
   (D) 176
   (E) 184

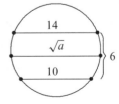

29. For how many three-element sets of positive integers $\{a, b, c\}$ is it true that $a \times b \times c = 2310$?

   (A) 32 (B) 36 (C) 40 (D) 43 (E) 45

30. A large cube is formed by stacking 27 unit cubes. A plane is perpendicular to one of the internal diagonals of the large cube and bisects that diagonal. The number of unit cubes that the plane intersects is

   (A) 16 (B) 17 (C) 18 (D) 19 (E) 20

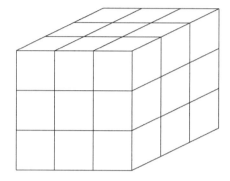

The 1995 AHSME was distributed to 5495 schools. It was written by 255,967 students. There were 8 perfect papers and 17,179 national Honor Roll students. The following table lists the percent of Honor Roll students who gave each answer to each question. The correct answer is the one in the first column.

ANSWER
#1:	(B) 99.67	(A) 0.07	(C) 0.15	(D) 0.05	(E) 0.05
#2:	(D) 98.17	(A) 0.10	(B) 0.86	(C) 0.47	(E) 0.36
#3:	(B) 99.27	(A) 0.16	(C) 0.44	(D) 0.02	(E) 0.01
#4:	(B) 97.36	(A) 0.33	(C) 0.03	(D) 1.62	(E) 0.17
#5:	(C) 96.53	(A) 0.34	(B) 2.77	(D) 0.19	(E) 0.10
#6:	(C) 98.95	(A) 0.02	(B) 0.06	(D) 0.26	(E) 0.21
#7:	(C) 93.38	(A) 4.31	(B) 0.85	(D) 0.16	(E) 0.19
#8:	(C) 95.88	(A) 0.90	(B) 2.53	(D) 0.05	(E) 0.06
#9:	(D) 96.56	(A) 0.00	(B) 0.24	(C) 2.27	(E) 0.22
#10:	(E) 97.94	(A) 0.09	(B) 0.10	(C) 0.10	(D) 0.13
#11:	(C) 67.06	(A) 0.51	(B) 1.38	(D) 6.16	(E) 0.81
#12:	(D) 51.56	(A) 0.51	(B) 1.34	(C) 0.80	(E) 0.69
#13:	(C) 71.09	(A) 0.08	(B) 0.32	(D) 0.27	(E) 5.39
#14:	(E) 95.67	(A) 0.01	(B) 0.15	(C) 0.03	(D) 0.19
#15:	(D) 85.31	(A) 2.35	(B) 7.12	(C) 0.29	(E) 1.38
#16:	(E) 35.68	(A) 0.73	(B) 5.01	(C) 1.18	(D) 49.13
#17:	(E) 40.35	(A) 2.55	(B) 12.81	(C) 1.26	(D) 0.90
#18:	(D) 56.94	(A) 2.94	(B) 4.83	(C) 4.33	(E) 1.25
#19:	(C) 54.33	(A) 1.15	(B) 3.88	(D) 4.46	(E) 2.18
#20:	(B) 37.92	(A) 1.57	(C) 1.68	(D) 1.64	(E) 1.21
#21:	(E) 4.79	(A) 26.96	(B) 9.36	(C) 5.87	(D) 2.75
#22:	(E) 48.89	(A) 0.22	(B) 0.27	(C) 0.24	(D) 0.39
#23:	(D) 25.54	(A) 0.73	(B) 22.41	(C) 3.41	(E) 3.12
#24:	(A) 35.76	(B) 0.51	(C) 0.34	(D) 0.26	(E) 0.57
#25:	(B) 16.05	(A) 1.96	(C) 2.90	(D) 1.36	(E) 1.14
#26:	(C) 15.73	(A) 0.21	(B) 0.24	(D) 0.41	(E) 1.61
#27:	(E) 13.37	(A) 0.55	(B) 0.59	(C) 1.66	(D) 0.59
#28:	(E) 4.00	(A) 3.69	(B) 0.39	(C) 0.47	(D) 0.58
#29:	(C) 6.69	(A) 1.26	(B) 2.26	(D) 0.73	(E) 2.54
#30:	(D) 4.13	(A) 0.68	(B) 1.88	(C) 2.01	(E) 0.17

47th AHSME, 1996

1. The addition below is incorrect. What is the largest digit that can be changed to make the addition correct?

    $$\begin{array}{r} 6\ 4\ 1 \\ 8\ 5\ 2 \\ +\ 9\ 7\ 3 \\ \hline 2\ 4\ 5\ 6 \end{array}$$

    (A) 4   **(B)** 5   **(C)** 6   **(D)** 7   **(E)** 8

2. Each day Walter gets $3 for doing his chores or $5 for doing them exceptionally well. After 10 days of doing his chores daily, Walter has received a total of $36. On how many days did Walter do them exceptionally well?

    (A) 3   **(B)** 4   **(C)** 5   **(D)** 6   **(E)** 7

3. $\dfrac{(3!)!}{3!} =$

    (A) 1   **(B)** 2   **(C)** 6   **(D)** 40   **(E)** 120

4. Six numbers from a list of nine integers are 7, 8, 3, 5, 9, and 5. The largest possible value of the median of all nine numbers in this list is

    (A) 5   **(B)** 6   **(C)** 7   **(D)** 8   **(E)** 9

5. Given that $0 < a < b < c < d$, which of the following is the largest?

(A) $\dfrac{a+b}{c+d}$ (B) $\dfrac{a+d}{b+c}$ (C) $\dfrac{b+c}{a+d}$

(D) $\dfrac{b+d}{a+c}$ (E) $\dfrac{c+d}{a+b}$

6. If $f(x) = x^{(x+1)}(x+2)^{(x+3)}$, then $f(0)+f(-1)+f(-2)+f(-3) =$

(A) $-8/9$ (B) 0   (C) $8/9$ (D) 1   (E) $10/9$

7. A father takes his twins and a younger child out to dinner on the twins' birthday. The restaurant charges $4.95 for the father and $0.45 for each year of a child's age, where age is defined as the age at the most recent birthday. If the bill is $9.45, which of the following could be the age of the youngest child?

(A) 1   (B) 2   (C) 3   (D) 4   (E) 5

8. If $3 = k \cdot 2^r$ and $15 = k \cdot 4^r$, then $r =$

(A) $-\log_2 5$ (B) $\log_5 2$ (C) $\log_{10} 5$

(D) $\log_2 5$ (E) $5/2$

9. Triangle PAB and square $ABCD$ are in perpendicular planes. Given that $PA = 3$, $PB = 4$, and $AB = 5$, what is PD?

(A) 5
(B) $\sqrt{34}$
(C) $\sqrt{41}$
(D) $2\sqrt{13}$
(E) 8

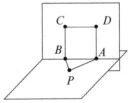

10. How many line segments have both their endpoints located at the vertices of a given cube?

(A) 12   (B) 15   (C) 24   (D) 28   (E) 56

11. Given a circle of radius 2, there are many line segments of length 2 that are tangent to the circle at their midpoints. Find the area of the region consisting of all such line segments.

   (A) $\pi/4$ (B) $4 - \pi$ (C) $\pi/2$ (D) π (E) 2π

12. A function f from the integers to the integers is defined as follows:
$$f(n) = \begin{cases} n + 3 & \text{if } n \text{ is odd,} \\ n/2 & \text{if } n \text{ is even.} \end{cases}$$
Suppose k is odd and $f(f(f(k))) = 27$. What is the sum of the digits of k?

   (A) 3   (B) 6   (C) 9   (D) 12   (E) 15

13. Sunny runs at a steady rate, and Moonbeam runs m times as fast, where m is a number greater than 1. If Moonbeam gives Sunny a head start of h meters, how many meters must Moonbeam run to overtake Sunny?

   (A) hm (B) $\dfrac{h}{h+m}$ (C) $\dfrac{h}{m-1}$ (D) $\dfrac{hm}{m-1}$ (E) $\dfrac{h+m}{m-1}$

14. Let $E(n)$ denote the sum of the even digits of n. For example, $E(5681) = 6 + 8 = 14$. Find $E(1) + E(2) + E(3) + \cdots + E(100)$.

   (A) 200   (B) 360   (C) 400   (D) 900   (E) 2250

15. Two opposite sides of a rectangle are each divided into n congruent segments, and the endpoints of one segment are joined to the center to form triangle A. The other sides are each divided into m congruent segments, and the endpoints of one of these segments are joined to the center to form triangle B. [See figure for $n = 5$, $m = 7$.] What is the ratio of the area of triangle A to the area of triangle B?

   (A) 1
   (B) m/n
   (C) n/m
   (D) $2m/n$
   (E) $2n/m$

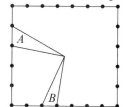

16. A fair standard six-sided die is tossed three times. Given that the sum of the first two tosses equals the third, what is the probability that at least one "2" is tossed?

   (A) $\dfrac{1}{6}$ (B) $\dfrac{91}{216}$ (C) $\dfrac{1}{2}$ (D) $\dfrac{8}{15}$ (E) $\dfrac{7}{12}$

17. In rectangle $ABCD$, angle C is trisected by $\overline{CF}$ and $\overline{CE}$, where E is on $\overline{AB}$, F is on $\overline{AD}$, $BE = 6$, and $AF = 2$. Which of the following is closest to the area of the rectangle $ABCD$?

   (A) 110
   (B) 120
   (C) 130
   (D) 140
   (E) 150

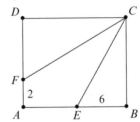

18. A circle of radius 2 has center at $(2, 0)$. A circle of radius 1 has center at $(5, 0)$. A line is tangent to the two circles at points in the first quadrant. Which of the following is closest to the y-intercept of the line?

   (A) $\sqrt{2}/4$ (B) $8/3$ (C) $1 + \sqrt{3}$ (D) $2\sqrt{2}$ (E) 3

19. The midpoints of the sides of a regular hexagon $ABCDEF$ are joined to form a smaller hexagon. What fraction of the area of $ABCDEF$ is enclosed by the smaller hexagon?

   (A) $1/2$
   (B) $\sqrt{3}/3$
   (C) $2/3$
   (D) $3/4$
   (E) $\sqrt{3}/2$

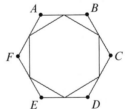

20. In the xy-plane, what is the length of the shortest path from $(0, 0)$ to $(12, 16)$ that does not go inside the circle $(x - 6)^2 + (y - 8)^2 = 25$?

   (A) $10\sqrt{3}$ (B) $10\sqrt{5}$ (C) $10\sqrt{3} + \dfrac{5\pi}{3}$
   (D) $40\dfrac{\sqrt{3}}{3}$ (E) $10 + 5\pi$

21. Triangles ABC and ABD are isosceles with $AB = AC = BD$, and $\overline{BD}$ intersects $\overline{AC}$ at E. If $\overline{BD} \perp \overline{AC}$, then $\angle C + \angle D$ is

   (A) 115°
   (B) 120°
   (C) 130°
   (D) 135°
   (E) not uniquely determined

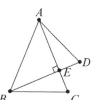

22. Four distinct points, A, B, C, and D, are to be selected from 1996 points evenly spaced around a circle. All quadruples are equally likely to be chosen. What is the probability that the chord $\overline{AB}$ intersects the chord $\overline{CD}$?

   (A) $\dfrac{1}{4}$ (B) $\dfrac{1}{3}$ (C) $\dfrac{1}{2}$ (D) $\dfrac{2}{3}$ (E) $\dfrac{3}{4}$

23. The sum of the lengths of the twelve edges of a rectangular box is 140, and the distance from one corner of the box to the farthest corner is 21. The total surface area of the box is

   (A) 776   (B) 784   (C) 798   (D) 800   (E) 812

24. The sequence

   $$1, 2, 1, 2, 2, 1, 2, 2, 2, 1, 2, 2, 2, 2, 1, 2, 2, 2, 2, 2, 1, 2, \ldots$$

   consists of 1's separated by blocks of 2's with n 2's in the nth block. The sum of the first 1234 terms of this sequence is

   (A) 1996   (B) 2419   (C) 2429   (D) 2439   (E) 2449

25. Given that $x^2 + y^2 = 14x + 6y + 6$, what is the largest possible value that $3x + 4y$ can have?

   (A) 72   (B) 73   (C) 74   (D) 75   (E) 76

26. An urn contains marbles of four colors: red, white, blue, and green. When four marbles are drawn without replacement, the following events are equally likely:
   (a) the selection of four red marbles;
   (b) the selection of one white and three red marbles;

(c) the selection of one white, one blue, and two red marbles; and
(d) the selection of one marble of each color.

What is the smallest number of marbles satisfying the given condition?

(A) 19   (B) 21   (C) 46   (D) 69   (E) more than 69

27. Consider two solid spherical balls, one centered at $(0, 0, \frac{21}{2})$ with radius 6, and the other centered at $(0, 0, 1)$ with radius $\frac{9}{2}$. How many points (x, y, z) with only integer coordinates (lattice points) are there in the intersection of the balls?

(A) 7   (B) 9   (C) 11   (D) 13   (E) 15

28. On a $4 \times 4 \times 3$ rectangular parallelepiped, vertices A, B, and C are adjacent to vertex D. The perpendicular distance from D to the plane containing A, B, and C is closest to

(A) 1.6   (B) 1.9   (C) 2.1   (D) 2.7   (E) 2.9

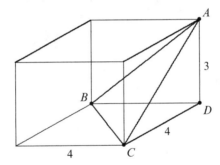

The 1996 AHSME was distributed to 5551 schools. It was written by 255,967 students. There were six perfect papers and 9,557 national Honor Roll students. The following table lists the percent of Honor Roll students who gave each answer to each question. The correct answer is the one in the first column.

```
ANSWER
  #1:  (D) 98.22   (A) 0.21   (B) 0.67    (C) 0.37    (E) 0.17
  #2:  (A) 99.08   (B) 0.08   (C) 0.09    (D) 0.36    (E) 0.30
  #3:  (E) 99.03   (A) 0.06   (B) 0.20    (C) 0.08    (D) 0.03
  #4:  (D) 87.86   (A) 1.18   (B) 1.01    (C) 2.58    (E) 1.55
  #5:  (E) 99.74   (A) 0.12   (B) 0.01    (C) 0.00    (D) 0.06
  #6:  (E) 89.07   (A) 2.25   (B) 0.19    (C) 7.46    (D) 0.15
  #7:  (B) 97.54   (A) 0.51   (C) 0.25    (D) 0.57    (E) 0.58
  #8:  (D) 82.78   (A) 0.70   (B) 0.63    (C) 0.15    (E) 0.09
  #9:  (B) 92.84   (A) 0.24   (C) 1.45    (D) 0.18    (E) 0.08
 #10:  (D) 89.33   (A) 3.44   (B) 0.54    (C) 1.62    (E) 0.89
 #11:  (D) 53.74   (A) 0.30   (B) 0.59    (C) 0.30    (E) 1.26
 #12:  (B) 80.76   (A) 5.38   (C) 9.38    (D) 0.30    (E) 0.52
 #13:  (D) 62.40   (A) 2.08   (B) 0.14    (C) 12.02   (E) 0.50
 #14:  (C) 91.45   (A) 0.70   (B) 1.37    (D) 0.19    (E) 0.18
 #15:  (B) 64.95   (A) 2.27   (C) 8.43    (D) 0.38    (E) 0.39
 #16:  (D) 60.03   (A) 0.69   (B) 0.66    (C) 2.95    (E) 0.91
 #17:  (E) 72.01   (A) 2.90   (B) 2.21    (C) 0.74    (D) 0.53
 #18:  (D) 15.62   (A) 0.50   (B) 24.84   (C) 2.84    (E) 3.39
 #19:  (D) 65.07   (A) 1.22   (B) 0.53    (C) 1.31    (E) 3.89
 #20:  (C) 16.07   (A) 0.77   (B) 2.45    (D) 2.54    (E) 8.84
 #21:  (D) 52.10   (A) 0.18   (B) 1.49    (C) 0.54    (E) 4.16
 #22:  (B) 13.93   (A) 2.29   (C) 5.77    (D) 0.75    (E) 0.49
 #23:  (B) 31.57   (A) 0.24   (C) 0.20    (D) 0.22    (E) 0.31
 #24:  (B) 35.74   (A) 0.51   (C) 0.45    (D) 0.87    (E) 0.64
 #25:  (B) 8.48    (A) 1.10   (C) 0.37    (D) 0.41    (E) 0.25
 #26:  (B) 2.62    (A) 0.25   (C) 2.68    (D) 0.31    (E) 0.88
 #27:  (D) 2.11    (A) 0.25   (B) 0.89    (C) 0.20    (E) 0.08
 #28:  (C) 16.18   (A) 0.97   (B) 1.74    (D) 2.20    (E) 0.84
 #29:  (C) 5.01    (A) 2.18   (B) 0.87    (D) 2.90    (E) 0.36
 #30:  (E) 0.23    (A) 0.15   (B) 0.17    (C) 0.12    (D) 0.21
```

48th AHSME, 1997

1. If a and b are digits for which

    $$\begin{array}{r} 2\,a \\ \times\ b\,3 \\ \hline 6\,9 \\ 9\,2 \\ \hline 9\,8\,9 \end{array}$$

    then a + b =

    (A) 3 (B) 4 (C) 7 (D) 9 (E) 12

2. The adjacent sides of the decagon shown meet at right angles. What is its perimeter?

    (A) 22
    (B) 32
    (C) 34
    (D) 44
    (E) 50

    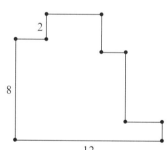

3. If x, y, and z are real numbers such that
    $$(x-3)^2 + (y-4)^2 + (z-5)^2 = 0,$$
    then $x + y + z =$

    (A) −12 (B) 0 (C) 8 (D) 12 (E) 50

4. If a is 50% larger than c, and b is 25% larger than c, then a is what percent larger than b?

    (A) 20%   (B) 25%   (C) 50%   (D) 100%   (E) 200%

5. A rectangle with perimeter 176 is divided into five congruent rectangles as shown in the diagram. What is the perimeter of one of the five congruent rectangles?

    (A) 35.2
    (B) 76
    (C) 80
    (D) 84
    (E) 86

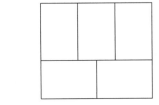

6. Consider the sequence
$$1, -2, 3, -4, 5, -6, \ldots,$$
whose nth term is $(-1)^{n+1} \cdot n$. What is the average of the first 200 terms of the sequence?

    (A) -1 (B) -0.5 (C) 0   (D) 0.5   (E) 1

7. The sum of seven integers is -1. What is the maximum number of the seven integers that can be larger than 13?

    (A) 1   (B) 4   (C) 5   (D) 6   (E) 7

8. Mientka Publishing Company prices its bestseller *Where's Walter?* as follows:
$$C(n) = \begin{cases} 12n, & \text{if } 1 \leq n \leq 24, \\ 11n, & \text{if } 25 \leq n \leq 48, \\ 10n, & \text{if } 49 \leq n, \end{cases}$$
where n is the number of books ordered, and $C(n)$ is the cost in dollars of n books. Notice that 25 books cost less than 24 books. For how many values of n is it cheaper to buy more than n books than to buy exactly n books?

    (A) 3   (B) 4   (C) 5   (D) 6   (E) 8

9. In the figure, $ABCD$ is a 2×2 square, E is the midpoint of $\overline{AD}$, and F is on $\overline{BE}$. If $\overline{CF}$ is perpendicular to $\overline{BE}$, then the area of quadrilateral $CDEF$ is

(A) 2

(B) $3 - \dfrac{\sqrt{3}}{2}$

(C) $\dfrac{11}{5}$

(D) $\sqrt{5}$

(E) $\dfrac{9}{4}$

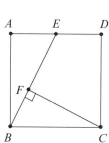

10. Two six-sided dice are fair in the sense that each face is equally likely to turn up. However, one of the dice has the 4 replaced by 3 and the other die has the 3 replaced by 4. When these dice are rolled, what is the probability that the sum is an odd number?

(A) $\dfrac{1}{3}$ (B) $\dfrac{4}{9}$ (C) $\dfrac{1}{2}$ (D) $\dfrac{5}{9}$ (E) $\dfrac{11}{18}$

11. In the sixth, seventh, eighth, and ninth basketball games of the season, a player scored 23, 14, 11, and 20 points, respectively. Her points-per-game average was higher after nine games than it was after the first five games. If her average after ten games was greater than 18, what is the least number of points she could have scored in the tenth game?

(A) 26 (B) 27 (C) 28 (D) 29 (E) 30

12. If m and b are real numbers and $mb > 0$, then the line whose equation is $y = mx + b$ <u>cannot</u> contain the point

(A) $(0, 1997)$ (B) $(0, -1997)$ (C) $(19, 97)$

(D) $(19, -97)$ (E) $(1997, 0)$

13. How many two-digit positive integers N have the property that the sum of N and the number obtained by reversing the order of the digits of N is a perfect square?

(A) 4 (B) 5 (C) 6 (D) 7 (E) 8

14. The number of geese in a flock increases so that the difference between the populations in year $n+2$ and year n is directly proportional to the population in year $n+1$. If the populations in the years 1994, 1995, and 1997 were 39, 60, and 123, respectively, then the population in 1996 was

    (A) 81   (B) 84   (C) 87   (D) 90   (E) 102

15. Medians BD and CE of triangle ABC are perpendicular, $BD = 8$, and $CE = 12$. The area of triangle ABC is

    (A) 24   (B) 32
    (C) 48   (D) 64
    (E) 96

    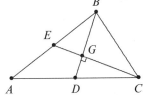

16. The three row sums and the three column sums of the array

    $$\begin{bmatrix} 4 & 9 & 2 \\ 8 & 1 & 6 \\ 3 & 5 & 7 \end{bmatrix}$$

    are the same. What is the least number of entries that must be altered to make all six sums different from one another?

    (A) 1   (B) 2   (C) 3   (D) 4   (E) 5

17. A line $x = k$ intersects the graph of $y = \log_5 x$ and the graph of $y = \log_5(x+4)$. The distance between the points of intersection is 0.5. Given that $k = a + \sqrt{b}$, where a and b are integers, what is $a + b$?

    (A) 6   (B) 7   (C) 8   (D) 9   (E) 10

18. A list of integers has mode 32 and mean 22. The smallest number in the list is 10. The median m of the list is a member of the list. If the list member m were replaced by $m + 10$, the mean and median of the new list would be 24 and $m + 10$, respectively. If m were instead replaced by $m - 8$, the median of the new list would be $m - 4$. What is m?

    (A) 16   (B) 17   (C) 18   (D) 19   (E) 20

19. A circle with center O is tangent to the coordinate axes and to the hypotenuse of the 30°-60°-90° triangle ABC as shown, where $AB = 1$. To the nearest hundredth, what is the radius of the circle?

(A) 2.18
(B) 2.24
(C) 2.31
(D) 2.37
(E) 2.41

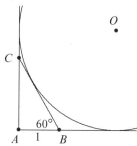

20. Which one of the following integers can be expressed as the sum of 100 consecutive positive integers?

(A) 1,627,384,950   (B) 2,345,678,910   (C) 3,579,111,300
(D) 4,692,581,470   (E) 5,815,937,260

21. For any positive integer n, let

$$f(n) = \begin{cases} \log_8 n, & \text{if } \log_8 n \text{ is rational,} \\ 0, & \text{otherwise.} \end{cases}$$

What is $\sum_{n=1}^{1997} f(n)$?

(A) $\log_8 2047$ (B) 6   (C) $\dfrac{55}{3}$ (D) $\dfrac{58}{3}$ (E) 585

22. Ashley, Betty, Carlos, Dick, and Elgin went shopping. Each had a whole number of dollars to spend, and together they had $56. The absolute difference between the amounts Ashley and Betty had to spend was $19. The absolute difference between the amounts Betty and Carlos had was $7, between Carlos and Dick was $5, between Dick and Elgin was $4, and between Elgin and Ashley was $11. How much did Elgin have?

(A) $6   (B) $7   (C) $8   (D) $9   (E) $10

23. In the figure, polygons A, E, and F are isosceles right triangles; B, C, and D are squares with sides of length 1; and G is an equilateral triangle. The figure can be folded along its edges to form a polyhedron having the polygons as faces. The volume of this polyhedron is

(A) 1/2

(B) 2/3

(C) 3/4

(D) 5/6

(E) 4/3

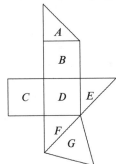

24. A *rising* number, such as 34689, is a positive integer each digit of which is larger than each of the digits to its left. There are $\binom{9}{5} = 126$ five-digit rising numbers. When these numbers are arranged from smallest to largest, the 97th number in the list does not contain the digit

(A) 4   (B) 5   (C) 6   (D) 7   (E) 8

25. Let $ABCD$ be a parallelogram and let $\overrightarrow{AA'}$, $\overrightarrow{BB'}$, $\overrightarrow{CC'}$, and $\overrightarrow{DD'}$ be parallel rays in space on the same side of the plane determined by $ABCD$. If $AA' = 10$, $BB' = 8$, $CC' = 18$, $DD' = 22$, and M and N are the midpoints of $\overline{A'C'}$ and $\overline{B'D'}$, respectively, then $MN =$

(A) 0   (B) 1   (C) 2   (D) 3   (E) 4

26. Triangle ABC and point P in the same plane are given. Point P is equidistant from A and B, angle APB is twice angle ACB, and $\overline{AC}$ intersects $\overline{BP}$ at point D. If $PB = 3$ and $PD = 2$, then $AD \cdot CD =$

(A) 5

(B) 6

(C) 7

(D) 8

(E) 9

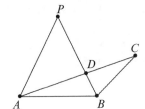

27. Consider those functions f that satisfy $f(x+4) + f(x-4) = f(x)$ for all real x. Any such function is periodic, and there is a least common positive period p for all of them. Find p.

    (A) 8   (B) 12   (C) 16   (D) 24   (E) 32

28. How many ordered triples of integers (a, b, c) satisfy
$$|a+b| + c = 19 \text{ and } ab + |c| = 97?$$

    (A) 0   (B) 4   (C) 6   (D) 10   (E) 12

29. Call a positive real number *special* if it has a decimal representation that consists entirely of digits 0 and 7. For example, $\frac{700}{99} = 7.\overline{07} = 7.070707\ldots$ and 77.007 are special numbers. What is the smallest n such that 1 can be written as a sum of n special numbers?

    (A) 7   (B) 8   (C) 9   (D) 10

    (E) The number 1 cannot be represented as a sum of finitely many special numbers.

30. For positive integers n, denote by $D(n)$ the number of pairs of different adjacent digits in the binary (base two) representation of n. For example, $D(3) = D(11_2) = 0$, $D(21) = D(10101_2) = 4$, and $D(97) = D(1100001_2) = 2$. For how many positive integers n less than or equal to 97 does $D(n) = 2$?

    (A) 16   (B) 20   (C) 26   (D) 30   (E) 35

The 1997 AHSME was distributed to 5635 schools. It was written by 268,952 students. There were no perfect papers and 6,589 national Honor Roll students. The following table lists the percent of Honor Roll students who gave each answer to each question. The correct answer is the one in the first column.

```
ANSWER
#1:  (C) 98.89  (A) 0.68  (B) 0.06  (D) 0.00  (E) 0.17
#2:  (D) 99.50  (A) 0.00  (B) 0.06  (C) 0.17  (E) 0.05
#3:  (D) 98.27  (A) 0.02  (B) 0.29  (C) 0.05  (E) 0.03
#4:  (A) 99.42  (B) 0.20  (C) 0.02  (D) 0.03  (E) 0.12
#5:  (C) 93.11  (A) 0.26  (B) 0.73  (D) 0.38  (E) 0.20
#6:  (B) 95.96  (A) 1.88  (C) 0.52  (D) 0.35  (E) 0.06
#7:  (D) 98.36  (A) 0.00  (B) 0.05  (C) 0.08  (E) 0.21
#8:  (D) 92.61  (A) 0.46  (B) 0.32  (C) 2.11  (E) 0.26
#9:  (C) 75.06  (A) 2.02  (B) 5.95  (D) 0.27  (E) 0.33
#10: (D) 89.44  (A) 0.24  (B) 3.49  (C) 2.44  (E) 0.55
#11: (D) 65.93  (A) 0.90  (B) 2.88  (C) 21.43 (E) 0.52
#12: (E) 94.25  (A) 0.83  (B) 0.85  (C) 0.17  (D) 0.88
#13: (E) 78.65  (A) 3.60  (B) 2.53  (C) 1.18  (D) 0.55
#14: (B) 82.85  (A) 0.58  (C) 1.26  (D) 0.26  (E) 1.05
#15: (D) 27.79  (A) 0.18  (B) 0.42  (C) 3.57  (E) 1.00
#16: (D) 50.95  (A) 0.09  (B) 3.37  (C) 19.12 (E) 5.72
#17: (A) 41.19  (B) 0.30  (C) 0.21  (D) 0.47  (E) 1.18
#18: (E) 42.95  (A) 0.39  (B) 0.12  (C) 0.30  (D) 0.38
#19: (D) 15.12  (A) 0.30  (B) 0.41  (C) 0.99  (E) 0.56
#20: (A) 63.99  (B) 0.11  (C) 4.87  (D) 0.06  (E) 0.15
#21: (C) 32.33  (A) 0.35  (B) 8.82  (D) 0.38  (E) 2.32
#22: (E) 58.48  (A) 0.68  (B) 0.18  (C) 0.61  (D) 0.96
#23: (D) 19.96  (A) 1.37  (B) 2.44  (C) 6.03  (E) 0.35
#24: (B) 14.14  (A) 3.51  (C) 1.21  (D) 1.12  (E) 1.08
#25: (B) 9.14   (A) 1.50  (C) 0.46  (D) 0.94  (E) 0.15
#26: (A) 6.09   (B) 0.93  (C) 0.12  (D) 0.29  (E) 0.14
#27: (D) 4.74   (A) 2.22  (B) 0.67  (C) 2.17  (E) 0.32
#28: (E) 0.49   (A) 1.68  (B) 1.53  (C) 0.44  (D) 0.14
#29: (B) 1.40   (A) 0.33  (C) 0.46  (D) 0.68  (E) 11.78
#30: (C) 10.24  (A) 0.90  (B) 0.94  (D) 0.82  (E) 1.35
```

49th AHSME, 1998

1.

   $$\begin{array}{c}A\\ \boxed{\begin{array}{ccc}&1&\\4&&6\\&9&\end{array}}\end{array}\quad \begin{array}{c}B\\ \boxed{\begin{array}{ccc}&0&\\1&&3\\&6&\end{array}}\end{array}\quad \begin{array}{c}C\\ \boxed{\begin{array}{ccc}&8&\\3&&5\\&2&\end{array}}\end{array}\quad \begin{array}{c}D\\ \boxed{\begin{array}{ccc}&5&\\7&&4\\&8&\end{array}}\end{array}\quad \begin{array}{c}E\\ \boxed{\begin{array}{ccc}&2&\\9&&7\\&0&\end{array}}\end{array}$$

   Each of the sides of five congruent rectangles is labeled with an integer, as shown above. These five rectangles are placed, without rotating or reflecting, in positions I through V so that the labels on coincident sides are equal.

I	II	III
IV	V	

   Which of the rectangles is in position I?

   (A) A **(B)** B **(C)** C **(D)** D **(E)** E

2. Letters A, B, C, and D represent four different digits selected from $0, 1, 2, \ldots, 9$. If $(A + B)/(C + D)$ is an integer that is as large as possible, what is the value of $A + B$?

   (A) 13   **(B)** 14   **(C)** 15   **(D)** 16   **(E)** 17

3. If a, b, and c are digits for which

   ```
     7 a 2
   - 4 8 b
     c 7 3
   ```

   then a + b + c =

   (A) 14   **(B)** 15   **(C)** 16   **(D)** 17   **(E)** 18

4. Define $[a, b, c]$ to mean $\dfrac{a+b}{c}$, where $c \neq 0$. What is the value of
$$[[60, 30, 90], [2, 1, 3], [10, 5, 15]]?$$

   (A) 0   (B) 0.5   (C) 1   (D) 1.5   (E) 2

5. If $2^{1998} - 2^{1997} - 2^{1996} + 2^{1995} = k \cdot 2^{1995}$, what is the value of k?

   (A) 1   (B) 2   (C) 3   (D) 4   (E) 5

6. If 1998 is written as a product of two positive integers whose difference is as small as possible, then the difference is

   (A) 8   (B) 15   (C) 17   (D) 47   (E) 93

7. If $N > 1$, then $\sqrt[3]{N \sqrt[3]{N \sqrt[3]{N}}} =$

   (A) $N^{\frac{1}{27}}$ (B) $N^{\frac{1}{9}}$ (C) $N^{\frac{1}{3}}$ (D) $N^{\frac{13}{27}}$ (E) N

8. A square with sides of length 1 is divided into two congruent trapezoids and a pentagon, which have equal areas, by joining the center of the square with points on three of the sides, as shown. Find x, the length of the longer parallel side of each trapezoid.

   (A) 3/5   (B) 2/3
   (C) 3/4   (D) 5/6
   (E) 7/8

9. A speaker talked for sixty minutes to a full auditorium. Twenty percent of the audience heard the entire talk and ten percent slept through the entire talk. Half of the remainder heard one third of the talk and the other half heard two thirds of the talk. What was the average number of minutes of the talk heard by members of the audience?

   (A) 24   (B) 27   (C) 30   (D) 33   (E) 36

10. A large square is divided into a small square surrounded by four congruent rectangles as shown. The perimeter of each of the congruent rectangles is 14. What is the area of the large square?

    (A) 49   (B) 64   (C) 100   (D) 121   (E) 196

11. Let R be a rectangle. How many circles in the plane of R have a diameter both of whose endpoints are vertices of R?

    (A) 1   (B) 2   (C) 4   (D) 5   (E) 6

12. How many different prime numbers are factors of N if
    $$\log_2(\log_3(\log_5(\log_7 N))) = 11?$$

    (A) 1   (B) 2   (C) 3   (D) 4   (E) 7

13. Walter rolls four standard six-sided dice and finds that the product of the numbers on the upper faces is 144. Which of the following could **not** be the sum of the upper four faces?

    (A) 14   (B) 15   (C) 16   (D) 17   (E) 18

14. A parabola has vertex at $(4, -5)$ and has two x-intercepts, one positive and one negative. If this parabola is the graph of $y = ax^2 + bx + c$, which of $a, b,$ and c must be positive?

    (A) only a (B) only b (C) only c (D) a and b only (E) none

15. A regular hexagon and an equilateral triangle have equal areas. What is the ratio of the length of a side of the triangle to the length of a side of the hexagon?

    (A) $\sqrt{3}$ (B) 2   (C) $\sqrt{6}$ (D) 3   (E) 6

16. The figure shown is the union of a circle and two semicircles of diameters a and b, all of whose centers are collinear. The ratio of the area of the shaded region to that of the unshaded region is

    (A) $\sqrt{\dfrac{a}{b}}$ (B) $\dfrac{a}{b}$

    (C) $\dfrac{a^2}{b^2}$ (D) $\dfrac{a+b}{2b}$

    (E) $\dfrac{a^2 + 2ab}{b^2 + 2ab}$

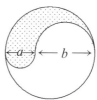

17. Let $f(x)$ be a function with the two properties:

    (a) for any two real numbers x and y, $f(x+y) = x + f(y)$, and
    (b) $f(0) = 2$.

    What is the value of $f(1998)$?

    (A) 0   (B) 2   (C) 1996   (D) 1998   (E) 2000

18. A right circular cone of volume A, a right circular cylinder of volume M, and a sphere of volume C all have the same radius, and the common height of the cone and the cylinder is equal to the diameter of the sphere. Then

    (A) $A - M + C = 0$ (B) $A + M = C$ (C) $2A = M + C$
    (D) $A^2 - M^2 + C^2 = 0$ (E) $2A + 2M = 3C$

19. How many triangles have area 10 and vertices at $(-5, 0), (5, 0)$, and $(5\cos\theta, 5\sin\theta)$ for some angle θ?

    (A) 0   (B) 2   (C) 4   (D) 6   (E) 8

20. Three cards, each with a positive integer written on it, are lying face-down on a table. Casey, Stacy, and Tracy are told that

    (a) the numbers are all different,
    (b) they sum to 13, and
    (c) they are in increasing order, left to right.

    First, Casey looks at the number on the leftmost card and says, "I don't have enough information to determine the other two numbers." Then Tracy looks at the number on the rightmost card and says, "I don't have enough information to determine the other two numbers." Finally, Stacy looks at the number on the middle card and says, "I don't have enough information to determine the other two numbers." Assume that each person knows that the other two reason perfectly and hears their comments. What number is on the middle card?

    (A) 2   (B) 3   (C) 4   (D) 5
    (E) There is not enough information to determine the number.

21. In an h-meter race, Sunny is exactly d meters ahead of Windy when Sunny finishes the race. The next time they race, Sunny sportingly starts d meters behind Windy, who is at the starting line. Both runners run at the same constant speed as they did in the first race. How many meters ahead is Sunny when Sunny finishes the second race?

    (A) $\dfrac{d}{h}$ (B) 0 (C) $\dfrac{d^2}{h}$ (D) $\dfrac{h^2}{d}$ (E) $\dfrac{d^2}{h-d}$

22. What is the value of the expression
    $$\frac{1}{\log_2 100!} + \frac{1}{\log_3 100!} + \frac{1}{\log_4 100!} + \cdots + \frac{1}{\log_{100} 100!}?$$

    (A) 0.01 (B) 0.1 (C) 1 (D) 2 (E) 10

23. The graphs of $x^2 + y^2 = 4 + 12x + 6y$ and $x^2 + y^2 = k + 4x + 12y$ intersect when k satisfies $a \leq k \leq b$, and for no other values of k. Find $b - a$.

    (A) 5 (B) 68 (C) 104 (D) 140 (E) 144

24. Call a 7-digit telephone number $d_1d_2d_3\text{-}d_4d_5d_6d_7$ *memorable* if the prefix sequence $d_1d_2d_3$ is exactly the same as either of the sequences $d_4d_5d_6$ or $d_5d_6d_7$ (possibly both). Assuming that each d_i can be any of the ten decimal digits $0, 1, 2, \ldots 9$, the number of different memorable telephone numbers is

    (A) 19,810 (B) 19,910 (C) 19,990
    (D) 20,000 (E) 20,100

25. A piece of graph paper is folded once so that $(0,2)$ is matched with $(4,0)$, and $(7,3)$ is matched with (m, n). Find $m + n$.

    (A) 6.7 (B) 6.8 (C) 6.9 (D) 7.0 (E) 8.0

26. In quadrilateral $ABCD$, it is given that $\angle A = 120°$, angles B and D are right angles, $AB = 13$, and $AD = 46$. Then $AC =$

    (A) 60 (B) 62 (C) 64 (D) 65 (E) 72

27. A $9 \times 9 \times 9$ cube is composed of twenty-seven $3 \times 3 \times 3$ cubes. The big cube is 'tunneled' as follows: First, the six $3 \times 3 \times 3$ cubes which make up the center of each face as well as the center $3 \times 3 \times 3$ cube are removed as shown. Second, each of the twenty remaining $3 \times 3 \times 3$ cubes is diminished in the same way. That is, the center facial unit cubes as well as each center cube are removed. The surface area of the final figure is

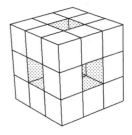

(A) 384 (B) 729 (C) 864 (D) 1024 (E) 1056

28. In triangle ABC, angle C is a right angle and $CB > CA$. Point D is located on $\overline{BC}$ so that angle CAD is twice angle DAB. If $AC/AD = 2/3$, then $CD/BD = m/n$, where m and n are relatively prime positive integers. Find $m + n$.

(A) 10 (B) 14 (C) 18 (D) 22 (E) 26

29. A point (x, y) in the plane is called a *lattice point* if both x and y are integers. The area of the largest square that contains exactly three lattice points in its interior is closest to

(A) 4.0 (B) 4.2 (C) 4.5 (D) 5.0 (E) 5.6

30. For each positive integer n, let
$$a_n = \frac{(n+9)!}{(n-1)!}$$
Let k denote the smallest positive integer for which the rightmost nonzero digit of a_k is odd. The rightmost nonzero digit of a_k is

(A) 1 (B) 3 (C) 5 (D) 7 (E) 9

The 1998 AHSME was distributed to 5354 schools. It was written by 243,996 students. There were no perfect papers and 8,643 national Honor Roll students. The following table lists the percent of Honor Roll students who gave each answer to each question. The correct answer is the one in the first column.

ANSWER
#1: (E) 95.96 (A) 0.38 (B) 0.24 (C) 0.14 (D) 1.38
#2: (E) 95.36 (A) 0.01 (B) 0.03 (C) 3.18 (D) 1.28
#3: (D) 98.53 (A) 0.06 (B) 0.20 (C) 0.61 (E) 0.45
#4: (E) 97.25 (A) 0.01 (B) 0.06 (C) 2.02 (D) 0.03
#5: (C) 96.86 (A) 0.34 (B) 0.76 (D) 0.42 (E) 0.23
#6: (C) 95.97 (A) 0.06 (B) 0.08 (D) 0.35 (E) 2.41
#7: (D) 93.83 (A) 5.04 (B) 0.15 (C) 0.07 (E) 0.22
#8: (D) 84.68 (A) 0.17 (B) 1.10 (C) 1.24 (E) 0.14
#9: (D) 97.27 (A) 0.07 (B) 0.27 (C) 0.52 (E) 0.35
#10: (A) 97.73 (B) 0.37 (C) 0.03 (D) 0.05 (E) 0.37
#11: (D) 43.06 (A) 3.41 (B) 0.79 (C) 1.21 (E) 39.33
#12: (A) 56.03 (B) 1.53 (C) 0.34 (D) 8.17 (E) 0.62
#13: (E) 90.39 (A) 0.86 (B) 0.78 (C) 0.58 (D) 2.05
#14: (A) 77.24 (B) 1.48 (C) 1.13 (D) 3.26 (E) 1.25
#15: (C) 69.96 (A) 2.49 (B) 1.08 (D) 0.96 (E) 6.05
#16: (B) 55.54 (A) 0.22 (C) 1.72 (D) 0.32 (E) 9.67
#17: (E) 85.17 (A) 0.08 (B) 0.19 (C) 0.25 (D) 0.61
#18: (A) 65.21 (B) 3.08 (C) 0.28 (D) 0.15 (E) 0.38
#19: (C) 40.83 (A) 1.90 (B) 9.71 (D) 0.67 (E) 0.40
#20: (C) 26.23 (A) 0.25 (B) 1.63 (D) 2.59 (E) 20.07
#21: (C) 42.57 (A) 1.39 (B) 4.30 (D) 0.45 (E) 2.02
#22: (C) 30.61 (A) 0.69 (B) 0.90 (D) 0.57 (E) 1.31
#23: (D) 11.71 (A) 0.36 (B) 0.75 (C) 0.60 (E) 0.49
#24: (C) 21.38 (A) 1.17 (B) 0.58 (D) 7.21 (E) 0.76
#25: (B) 21.65 (A) 0.57 (C) 0.32 (D) 2.38 (E) 3.39
#26: (B) 16.01 (A) 0.42 (C) 0.43 (D) 0.28 (E) 0.29
#27: (E) 12.48 (A) 2.21 (B) 0.79 (C) 3.46 (D) 0.71
#28: (B) 6.91   (A) 0.23 (C) 0.44 (D) 0.21 (E) 0.24
#29: (D) 1.30   (A) 1.18 (B) 0.45 (C) 4.25 (E) 0.22
#30: (E) 2.38   (A) 1.98 (B) 0.80 (C) 1.52 (D) 0.68

50th AHSME, 1999

1. $1 - 2 + 3 - 4 + \cdots - 98 + 99 =$

   (A) -50 **(B)** -49 **(C)** 0 **(D)** 49 **(E)** 50

2. Which one of the following statements is false?

   (A) All equilateral triangles are congruent to each other.

   (B) All equilateral triangles are convex.

   (C) All equilateral triangles are equiangular.

   (D) All equilateral triangles are regular polygons.

   (E) All equilateral triangles are similar to each other.

3. The number halfway between $1/8$ and $1/10$ is

   (A) $\dfrac{1}{80}$ **(B)** $\dfrac{1}{40}$ **(C)** $\dfrac{1}{18}$ **(D)** $\dfrac{1}{9}$ **(E)** $\dfrac{9}{80}$

4. Find the sum of all prime numbers between 1 and 100 that are simultaneously 1 greater than a multiple of 4 and 1 less than a multiple of 5.

   (A) 118 **(B)** 137 **(C)** 158 **(D)** 187 **(E)** 245

5. The marked price of a book was 30% less than the suggested retail price. Alice purchased the book for half the marked price at a Fiftieth Anniversary sale. What percent of the suggested retail price did Alice pay?

   (A) 25% **(B)** 30% **(C)** 35% **(D)** 60% **(E)** 65%

6. What is the sum of the digits of the decimal form of the product $2^{1999} \cdot 5^{2001}$?

   (A) 2   (B) 4   (C) 5   (D) 7   (E) 10

7. What is the largest number of acute angles that a convex hexagon can have?

   (A) 2   (B) 3   (C) 4   (D) 5   (E) 6

8. At the end of 1994 Walter was half as old as his grandmother. The sum of the years in which they were born is 3838. How old will Walter be at the end of 1999?

   (A) 48   (B) 49   (C) 53   (D) 55   (E) 101

9. Before Ashley started a two-hour drive, her car's odometer reading was 27972, a palindrome. (A palindrome is a number that reads the same way from left to right as it does from right to left.) At her destination, the odometer reading was another palindrome. If Ashley never exceeded the speed limit of 75 miles per hour, which of the following was her average speed?

   (A) 50   (B) 55   (C) 60   (D) 65   (E) 70

10. A sealed envelope contains a card with a single digit on it. Three of the following statements are true, and the other is false.
    I. The digit is 1.
    II. The digit is not 2.
    III. The digit is 3.
    IV. The digit is not 4.
    Which one of the following must necessarily be correct?

    (A) I is true.   (B) I is false.   (C) II is true.

    (D) III is true.   (E) IV is false.

11. The student lockers at Olympic High are numbered consecutively beginning with locker number 1. The plastic digits used to number the lockers cost two cents apiece. Thus, it costs two cents to label locker number 9 and four cents to label locker number 10. If it costs $137.94 to label all the lockers, how many lockers are there at the school?

    (A) 2001   (B) 2010   (C) 2100   (D) 2726   (E) 6897

12. What is the maximum number of points of intersection of the graphs of two different fourth degree polynomial functions $y = p(x)$ and $y = q(x)$, each with leading coefficient 1?

    (A) 1   (B) 2   (C) 3   (D) 4   (E) 8

13. Define a sequence of real numbers $a_1, a_2, a_3, \ldots$ by $a_1 = 1$ and $a_{n+1}^3 = 99 a_n^3$ for all $n \geq 1$. Then a_{100} equals

    (A) 33^{33} (B) 33^{99} (C) 99^{33}
    (D) 99^{99} (E) none of these

14. Four girls — Mary, Alina, Tina, and Hanna — sang songs in a concert as trios, with one girl sitting out each time. Hanna sang seven songs, which was more than any other girl, and Mary sang four songs, which was fewer than any other girl. How many songs did these trios sing?

    (A) 7   (B) 8   (C) 9   (D) 10   (E) 11

15. Let x be a real number such that $\sec x - \tan x = 2$. Then $\sec x + \tan x =$

    (A) 0.1   (B) 0.2   (C) 0.3   (D) 0.4   (E) 0.5

16. What is the radius of a circle inscribed in a rhombus with diagonals of length 10 and 24?

    (A) 4   (B) 58/13   (C) 60/13   (D) 5   (E) 6

17. Let $P(x)$ be a polynomial such that when $P(x)$ is divided by $x - 19$, the remainder is 99, and when $P(x)$ is divided by $x - 99$, the remainder is 19. What is the remainder when $P(x)$ is divided by $(x-19)(x-99)$?

    (A) $-x + 80$ (B) $x + 80$ (C) $-x + 118$
    (D) $x + 118$ (E) 0

18. How many zeros does $f(x) = \cos(\log(x))$ have on the interval $0 < x < 1$?

    (A) 0   (B) 1   (C) 2   (D) 10   (E) infinitely many

19. Consider all triangles ABC satisfying the following conditions: $AB = AC$, D is a point on $\overline{AC}$ for which $\overline{BD} \perp \overline{AC}$, AD, and CD are integers, and $BD^2 = 57$. Among all such triangles, the smallest possible value of AC is

(A) 9 (B) 10 (C) 11

(D) 12 (E) 13

20. The sequence $a_1, a_2, a_3, \ldots$ satisfies $a_1 = 19$, $a_9 = 99$, and, for all $n \geq 3$, a_n is the arithmetic mean of the first $n - 1$ terms. Find a_2.

(A) 29 (B) 59 (C) 79 (D) 99 (E) 179

21. A circle is circumscribed about a triangle with sides 20, 21, and 29, thus dividing the interior of the circle into four regions. Let A, B, and C be the areas of the non-triangular regions, with C being the largest. Then

(A) $A + B = C$ (B) $A + B + 210 = C$ (C) $A^2 + B^2 = C^2$

(D) $20A + 21B = 29C$ (E) $\dfrac{1}{A^2} + \dfrac{1}{B^2} = \dfrac{1}{C^2}$

22. The graphs of $y = -|x - a| + b$ and $y = |x - c| + d$ intersect at points $(2, 5)$ and $(8, 3)$. Find $a + c$.

(A) 7 (B) 8 (C) 10 (D) 13 (E) 18

23. The equiangular convex hexagon $ABCDEF$ has $AB = 1$, $BC = 4$, $CD = 2$, and $DE = 4$. The area of the hexagon is

(A) $\dfrac{15}{2}\sqrt{3}$ (B) $9\sqrt{3}$ (C) 16 (D) $\dfrac{39}{4}\sqrt{3}$ (E) $\dfrac{43}{4}\sqrt{3}$

24. Five points on a circle are given. Four of the chords joining pairs of the five points are selected at random. What is the probability that the four chords form a convex quadrilateral?

(A) $\dfrac{1}{210}$ (B) $\dfrac{1}{105}$ (C) $\dfrac{1}{42}$ (D) $\dfrac{1}{15}$ (E) $\dfrac{1}{14}$

25. There are unique integers $a_2, a_3, a_4, a_5, a_6, a_7$ such that
$$\frac{5}{7} = \frac{a_2}{2!} + \frac{a_3}{3!} + \frac{a_4}{4!} + \frac{a_5}{5!} + \frac{a_6}{6!} + \frac{a_7}{7!},$$
where $0 \le a_i < i$ for $i = 2, 3, \ldots, 7$. Find $a_2 + a_3 + a_4 + a_5 + a_6 + a_7$.

(A) 8 (B) 9 (C) 10 (D) 11 (E) 12

26. Three non-overlapping regular plane polygons, at least two of which are congruent, all have sides of length 1. The polygons meet at a point A in such a way that the sum of the three interior angles at A is $360°$. Thus the three polygons form a new polygon with A as an interior point. What is the largest possible perimeter that this polygon can have?

(A) 12 (B) 14 (C) 18 (D) 21 (E) 24

27. In triangle ABC, $3 \sin A + 4 \cos B = 6$ and $4 \sin B + 3 \cos A = 1$. Then $\angle C$ in degrees is

(A) 30 (B) 60 (C) 90 (D) 120 (E) 150

28. Let $x_1, x_2, \ldots, x_n$ be a sequence of integers such that
    (i) $-1 \le x_i \le 2$, for $i = 1, 2, 3, \ldots, n$;
    (ii) $x_1 + x_2 + \cdots + x_n = 19$; and
    (iii) $x_1^2 + x_2^2 + \cdots + x_n^2 = 99$.
Let m and M be the minimal and maximal possible values of $x_1^3 + x_2^3 + \cdots + x_n^3$, respectively. Then $M/m =$

(A) 3 (B) 4 (C) 5 (D) 6 (E) 7

29. A tetrahedron with four equilateral triangular faces has a sphere inscribed within it and a sphere circumscribed about it. For each of the four faces, there is a sphere tangent externally to the face at its center and to the circumscribed sphere. A point P is selected at random inside the circumscribed sphere. The probability that P lies inside one of the five small spheres is closest to

(A) 0 (B) 0.1 (C) 0.2 (D) 0.3 (E) 0.4

30. The number of ordered pairs of integers (m, n) for which $mn \geq 0$ and
$$m^3 + n^3 + 99mn = 33^3$$
is equal to

(A) 2   (B) 3   (C) 33   (D) 35   (E) 99

The 1999 AHSME was distributed to 5635 schools. It was written by 240,897 students. There were no perfect papers and 6,327 national Honor Roll students. The following table lists the percent of Honor Roll students who gave each answer to each question. The correct answer is the one in the first column.

ANSWER

#1:	(E) 96.43	(A) 2.07	(B) 0.22	(C) 0.40	(D) 0.81
#2:	(A) 96.49	(B) 1.14	(C) 0.21	(D) 0.11	(E) 0.70
#3:	(E) 99.29	(A) 0.35	(B) 0.05	(C) 0.02	(D) 0.28
#4:	(A) 94.39	(B) 0.35	(C) 0.14	(D) 1.50	(E) 1.96
#5:	(C) 99.27	(A) 0.02	(B) 0.11	(D) 0.03	(E) 0.46
#6:	(D) 82.27	(A) 0.16	(B) 0.38	(C) 0.25	(E) 0.41
#7:	(B) 69.72	(A) 10.68	(C) 3.38	(D) 1.30	(E) 1.83
#8:	(D) 99.37	(A) 0.02	(B) 0.08	(C) 0.22	(E) 0.02
#9:	(D) 93.35	(A) 1.25	(B) 0.03	(C) 4.24	(E) 0.19
#10:	(C) 93.25	(A) 0.32	(B) 3.30	(D) 0.05	(E) 1.22
#11:	(A) 97.00	(B) 0.41	(C) 0.19	(D) 0.19	(E) 0.47
#12:	(C) 35.50	(A) 1.56	(B) 3.10	(D) 20.50	(E) 4.03
#13:	(C) 64.90	(A) 0.09	(B) 0.21	(D) 4.87	(E) 4.66
#14:	(A) 88.70	(B) 2.18	(C) 0.46	(D) 0.19	(E) 0.30
#15:	(E) 75.87	(A) 0.06	(B) 0.13	(C) 0.28	(D) 0.28
#16:	(C) 65.45	(A) 0.38	(B) 1.12	(D) 0.66	(E) 0.41
#17:	(C) 12.04	(A) 0.54	(B) 1.22	(D) 1.07	(E) 1.79
#18:	(E) 39.92	(A) 16.42	(B) 25.27	(C) 0.58	(D) 0.13
#19:	(C) 63.30	(A) 1.25	(B) 0.65	(D) 0.25	(E) 0.14
#20:	(E) 68.03	(A) 0.55	(B) 0.25	(C) 0.11	(D) 0.52
#21:	(B) 28.18	(A) 1.25	(C) 2.24	(D) 0.95	(E) 0.89
#22:	(C) 48.54	(A) 0.22	(B) 1.74	(D) 0.51	(E) 0.16
#23:	(E) 15.66	(A) 0.55	(B) 0.55	(C) 0.19	(D) 0.87
#24:	(B) 11.89	(A) 2.12	(C) 0.89	(D) 1.03	(E) 1.00
#25:	(B) 19.39	(A) 0.43	(C) 0.62	(D) 0.44	(E) 0.32
#26:	(D) 17.45	(A) 2.99	(B) 3.95	(C) 0.52	(E) 2.64
#27:	(A) 7.10	(B) 0.57	(C) 1.68	(D) 0.65	(E) 1.96
#28:	(E) 3.62	(A) 0.17	(B) 0.17	(C) 0.16	(D) 0.09
#29:	(C) 2.78	(A) 0.30	(B) 0.73	(D) 0.57	(E) 0.43
#30:	(D) 2.83	(A) 1.87	(B) 0.76	(C) 1.34	(E) 0.24

Sample AMC 10, 1999

1. The number halfway between $1/6$ and $1/4$ is

    (A) $\dfrac{1}{24}$ (B) $\dfrac{1}{5}$ (C) $\dfrac{2}{9}$ (D) $\dfrac{5}{24}$ (E) $\dfrac{3}{14}$

2. The marked price of a coat was 40% less than the suggested retail price. Alice purchased the coat for half the marked price at a Fiftieth Anniversary sale. What percent less than the suggested retail price did Alice pay?

    (A) 20%   (B) 30%   (C) 60%   (D) 70%   (E) 80%

3. The mean of three numbers is ten more than the least of the numbers and fifteen less than the greatest of the three. If the median of the three numbers is 5, then the sum of the three is

    (A) 5   (B) 20   (C) 25   (D) 30   (E) 36

4. 4 What is the largest number of obtuse angles that a quadrilateral can have?

    (A) 0   (B) 1   (C) 2   (D) 3   (E) 4

5. Consider the sequence
$$1, -2, 3, -4, 5, -6, \ldots,$$
whose n^{th} term is $(-1)^{n+1} \cdot n$. What is the average of the first 200 terms of the sequence?

    (A) -1 (B) -0.5 (C) 0   (D) 0.5   (E) 1

6. What is the sum of the digits of the decimal form of the product $2^{1999} \cdot 5^{2000}$?

   (A) 5 (B) 7 (C) 10 (D) 15 (E) 50

7. Find the sum of all prime numbers between 1 and 100 that are simultaneously one greater than a multiple of 5 and one less than a multiple of 6.

   (A) 52 (B) 82 (C) 123 (D) 143 (E) 214

8. Two rectangles, A, with vertices at $(-2, 0), (0, 0), (0, 4)$, and $(-2, 4)$, and B, with vertices at $(1, 0), (5, 0), (1, 12)$, and $(5, 12)$, are simultaneously bisected by a line in the plane. What is the slope of this line?

   (A) -4 (B) -1 (C) 0 (D) 1 (E) 2

9. A two-inch cube $(2 \times 2 \times 2)$ of silver weighs three pounds and is worth $200. How much is a three-inch cube of silver worth?

   (A) $300 (B) $375 (C) $450 (D) $560 (E) $675

10. The outside surface of a $4 \times 6 \times 8$ block of unit cubes is painted. How many unit cubes have exactly one face painted?

    (A) 88 (B) 140 (C) 144 (D) 192 (E) 208

11. The adjacent sides of the decagon shown meet at right angles. What is its perimeter?

    (A) 22
    (B) 32
    (C) 34
    (D) 44
    (E) 50

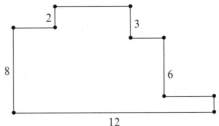

Sample AMC 10, 1999

12. Certain positive integers have these properties:

    I. the sum of the squares of their digits is 50;
    II. each digit is larger than the one to its left.

    The product of the digits of the largest integer with both properties is

    (A) 7   (B) 25   (C) 36   (D) 48   (E) 60

13. At the end of 1994 Walter was half as old as his grandmother. The sum of the years in which they were born is 3844. How old will Walter be at the end of 1999?

    (A) 48   (B) 49   (C) 53   (D) 55   (E) 101

14. All even numbers from 2 to 98 inclusive, except those ending in 0, are multiplied together. What is the rightmost digit (the units digit) of the product?

    (A) 0   (B) 2   (C) 4   (D) 6   (E) 8

15. How many three-element subsets of the set

    $$\{88, 95, 99, 132, 166, 173\}$$

    have the property that the sum of the three elements is even?

    (A) 6   (B) 8   (C) 10   (D) 12   (E) 24

16. A point is chosen at random from within a circular region. What is the probability that the point is closer to the center of the region than it is to the boundary of the region?

    (A) 1/4   (B) 1/3   (C) 1/2   (D) 2/3   (E) 3/4

17. It took 600 digits to label the pages of a book starting with page one. How many pages does the book have?

    (A) 136   (B) 137   (C) 236
    (D) 600   (E) none of A, B, C, or D

18. In the equation A + B + C + D + E = FG, where FG is the two-digit number whose value is 10F+G, and letters A, B, C, D, E, F, and G each represent **different** digits. If FG is as large as possible, what is the value of G?

    (A) 1   (B) 2   (C) 3   (D) 4   (E) 5

19. What is the maximum number of points of intersection of the graphs of two different cubic polynomial functions with leading coefficients 1?

   (A) 1   (B) 2   (C) 3   (D) 4   (E) 6

20. The graphs of $y = -|x - a| + b$ and $y = |x - c| + d$ intersect at points $(2, 5)$ and $(8, 3)$. Find $a + c$.

   (A) 5   (B) 7   (C) 8   (D) 10   (E) 13

21. A sealed envelope contains a card with a single digit on it. Three of the following statements are true, and the other is false.
   I. The digit is 1.
   II. The digit is 2.
   III. The digit is not 3.
   IV. The digit is not 4.
   Which one of the following must be correct?

   (A) I is false   (B) II is true   (C) II is false
   (D) III is false   (E) IV is true

22. A circle is circumscribed about a triangle with sides 3, 4, and 5, thus dividing the interior of the circle into four regions. Let A, B, and C be the areas of the non-triangular regions, with C being the largest. Then

   (A) $A + B = C$ (B) $A^2 + B^2 = C^2$ (C) $A + B + 6 = C$
   (D) $4A + 3B = 5C$ (E) $\dfrac{1}{A^2} + \dfrac{1}{B^2} = \dfrac{1}{C^2}$

23. The digits $2, 4, 5, 6, 8,$ and 9 can be distributed among the lettered squares in the array so that the sum of the entries on each of the rows and columns is the same number K. What is K?

   (A) 15
   (B) 16
   (C) 17
   (D) 19
   (E) 21

7	a	b	1
c			d
3	e	f	10

24. In a circle with center O, $\overline{OA}$ and $\overline{OB}$ are radii and $\angle AOB$ is a right angle. A semicircle is constructed using segment AB as its diameter as shown. The shaded portion of the semicircle outside circle O is called a *lune*. What is the ratio of the area of the lune to the area of the triangle?

(A) $\dfrac{\sqrt{2}}{\pi}$

(B) 1

(C) $\dfrac{\pi}{\sqrt{3}}$

(D) $\dfrac{\pi}{\sqrt{2}}$

(E) $\dfrac{2\pi}{3} - 1$

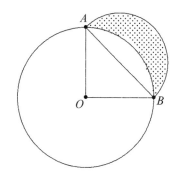

25. A regular hexagon and a regular pentagon have a common edge as shown. Find the measure of the angle BAC.

(A) 24°

(B) 30°

(C) 36°

(D) 45°

(E) 48°

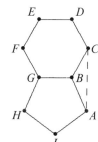

51st AMC 12, 2000

1. In the year 2001, the United States will host the International Mathematical Olympiad. Let $I, M,$ and O be distinct positive integers such that the product $I \cdot M \cdot O = 2001$. What is the largest possible value of the sum $I + M + O$?

    (A) 23 (B) 55 (C) 99 (D) 111 (E) 671

2. $2000(2000^{2000}) =$

    (A) 2000^{2001} (B) 4000^{2000} (C) 2000^{4000}
    (D) $4,000,000^{2000}$ (E) $2000^{4,000,000}$

3. Each day, Jenny ate 20% of the jellybeans that were in her jar at the beginning of that day. At the end of the second day, 32 remained. How many jellybeans were in the jar originally?

    (A) 40 (B) 50 (C) 55 (D) 60 (E) 75

4. The Fibonacci sequence $1, 1, 2, 3, 5, 8, 13, 21, \ldots$ starts with two 1s, and each term afterwards is the sum of its two predecessors. Which one of the ten digits is the last to appear in the units position of a number in the Fibonacci sequence?

    (A) 0 (B) 4 (C) 6 (D) 7 (E) 9

5. If $|x - 2| = p$, where $x < 2$, then $x - p =$

    (A) -2 (B) 2 (C) $2 - 2p$ (D) $2p - 2$ (E) $|2p - 2|$

6. Two different prime numbers between 4 and 18 are chosen. When their sum is subtracted from their product, which of the following numbers could be obtained?

   (A) 21   (B) 60   (C) 119   (D) 180   (E) 231

7. How many positive integers b have the property that $\log_b 729$ is a positive integer?

   (A) 0   (B) 1   (C) 2   (D) 3   (E) 4

8. Figures 0, 1, 2, and 3 consist of 1, 5, 13, and 25 nonoverlapping unit squares, respectively. If the pattern were continued, how many nonoverlapping unit squares would there be in figure 100?

   (A) 10401   (B) 19801   (C) 20201
   (D) 39801   (E) 40801

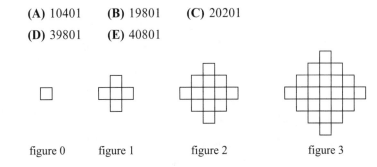

figure 0   figure 1   figure 2   figure 3

9. Mrs. Walter gave an exam in a mathematics class of five students. She entered the scores in random order into a spreadsheet, which recalculated the class average after each score was entered. Mrs. Walter noticed that after each score was entered, the average was always an integer. The scores (listed in ascending order) were 71, 76, 80, 82, and 91. What was the last score Mrs. Walter entered?

   (A) 71   (B) 76   (C) 80   (D) 82   (E) 91

10. The point $P = (1, 2, 3)$ is reflected in the xy-plane, then its image Q is rotated by $180°$ about the x-axis to produce R, and finally, R is translated by 5 units in the positive-y direction to produce S. What are the coordinates of S?

    (A) $(1, 7, -3)$ (B) $(-1, 7, -3)$ (C) $(-1, -2, 8)$
    (D) $(-1, 3, 3)$ (E) $(1, 3, 3)$

11. Two non-zero real numbers, a and b, satisfy $ab = a - b$. Which of the following is a possible value of $a/b + b/a - ab$?

   (A) -2 (B) $-\dfrac{1}{2}$ (C) $\dfrac{1}{3}$ (D) $\dfrac{1}{2}$ (E) 2

12. Let A, M, and C be nonnegative integers such that $A + M + C = 12$. What is the maximum value of $A \cdot M \cdot C + A \cdot M + M \cdot C + C \cdot A$?

   (A) 62   (B) 72   (C) 92   (D) 102   (E) 112

13. One morning each member of Angela's family drank an 8-ounce mixture of coffee with milk. The amounts of coffee and milk varied from cup to cup, but were never zero. Angela drank a quarter of the total amount of milk and a sixth of the total amount of coffee. How many people are in the family?

   (A) 3   (B) 4   (C) 5   (D) 6   (E) 7

14. When the mean, median, and mode of the list

$$10, 2, 5, 2, 4, 2, x$$

   are arranged in increasing order, they form a non-constant arithmetic progression. What is the sum of all possible real values of x?

   (A) 3   (B) 6   (C) 9   (D) 17   (E) 20

15. Let f be a function for which $f(x/3) = x^2 + x + 1$. Find the sum of all values of z for which $f(3z) = 7$.

   (A) $-1/3$ (B) $-1/9$ (C) 0 (D) $5/9$ (E) $5/3$

16. A checkerboard of 13 rows and 17 columns has a number written in each square, beginning in the upper left corner, so that the first row is numbered $1, 2, \ldots, 17$, the second row $18, 19, \ldots, 34$, and so on down the board. If the board is renumbered so that the left column, top to bottom, is $1, 2, \ldots, 13$, the second column $14, 15, \ldots, 26$ and so on across the board, some squares have the same numbers in both numbering systems. Find the sum of the numbers in these squares (under either system).

   (A) 222   (B) 333   (C) 444   (D) 555   (E) 666

17. A circle centered at O has radius 1 and contains the point A. Segment AB is tangent to the circle at A and $\angle AOB = \theta$. If point C lies on $\overline{OA}$ and $\overline{BC}$ bisects $\angle ABO$, then $OC =$

   (A) $\sec^2 \theta - \tan \theta$

   (B) $\dfrac{1}{2}$

   (C) $\dfrac{\cos^2 \theta}{1 + \sin \theta}$

   (D) $\dfrac{1}{1 + \sin \theta}$

   (E) $\dfrac{\sin \theta}{\cos^2 \theta}$

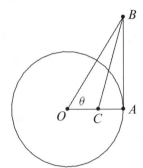

18. In year N, the 300th day of the year is a Tuesday. In year $N+1$, the 200th day is also a Tuesday. On what day of the week did the 100th day of year $N-1$ occur?

   (A) Thursday (B) Friday (C) Saturday
   (D) Sunday (E) Monday

19. In triangle ABC, $AB = 13$, $BC = 14$, and $AC = 15$. Let D denote the midpoint of $\overline{BC}$ and let E denote the intersection of $\overline{BC}$ with the bisector of angle BAC. Which of the following is closest to the area of the triangle ADE?

   (A) 2 (B) 2.5 (C) 3 (D) 3.5 (E) 4

20. If x, y, and z are positive numbers satisfying

   $$x + 1/y = 4, \quad y + 1/z = 1, \quad \text{and} \quad z + 1/x = 7/3,$$

   then $xyz =$

   (A) 2/3 (B) 1 (C) 4/3 (D) 2 (E) 7/3

21. Through a point on the hypotenuse of a right triangle, lines are drawn parallel to the legs of the triangle so that the triangle is divided into a square and two smaller right triangles. The area of one of the two small right triangles is m times the area of the square. The ratio of the area of the other small right triangle to the area of the square is

   (A) $\dfrac{1}{2m+1}$ (B) m (C) $1-m$ (D) $\dfrac{1}{4m}$ (E) $\dfrac{1}{8m^2}$

22. The graph below shows a portion of the curve defined by the quartic polynomial $P(x) = x^4 + ax^3 + bx^2 + cx + d$. Which of the following is the smallest?

   (A) $P(-1)$
   (B) The product of the zeros of P
   (C) The product of the non-real zeros of P
   (D) The sum of the coefficients of P
   (E) The sum of the real zeros of P

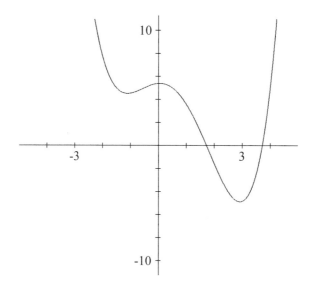

23. Professor Gamble buys a lottery ticket, which requires that he pick six different integers from 1 through 46, inclusive. He chooses his numbers so that the sum of the base-ten logarithms of his six numbers is an integer. It so happens that the integers on the winning ticket have the same property—the sum of the base-ten logarithms is an integer. What is the probability that Professor Gamble holds the winning ticket?

   (A) $1/5$ (B) $1/4$ (C) $1/3$ (D) $1/2$ (E) 1

24. If circular arcs AC and BC have centers at B and A, respectively, then there exists a circle tangent to both $\overset{\frown}{AC}$ and $\overset{\frown}{BC}$, and to $\overline{AB}$. If the length of $\overset{\frown}{BC}$ is 12, then the circumference of the circle is

(A) 24
(B) 25
(C) 26
(D) 27
(E) 28

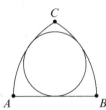

25. Eight congruent equilateral triangles, each of a different color, are used to construct a regular octahedron. How many distinguishable ways are there to construct the octahedron? (Two colored octahedrons are distinguishable if neither can be rotated to look just like the other.)

(A) 210
(B) 560
(C) 840
(D) 1260
(E) 1680

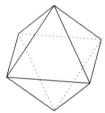

51st AMC 12, 2000

The 2000 AMC 10, AMC 12 was distributed to 5162 schools. The AMC 12 was written by 166,908 students. The AMC 10 was written by 106,994 students. Each contest had eight perfect papers. The number of AIME qualifiers through the AMC 12 was 9200. The following table lists the percent of AIME qualifiers on the AMC 12 who gave each answer to each question. The correct answer is the one in the first column.

ANSWER					
#1:	(E) 84.84	(A) 0.19	(B) 11.51	(C) 0.11	(D) 1.19
#2:	(A) 99.64	(B) 0.05	(C) 0.00	(D) 0.08	(E) 0.02
#3:	(B) 98.70	(A) 1.09	(C) 0.12	(D) 0.01	(E) 0.00
#4:	(C) 83.71	(A) 5.16	(B) 0.42	(D) 0.98	(E) 1.17
#5:	(C) 88.94	(A) 1.69	(B) 3.48	(D) 1.70	(E) 0.52
#6:	(C) 99.01	(A) 0.16	(B) 0.15	(D) 0.05	(E) 0.05
#7:	(E) 70.59	(A) 0.34	(B) 1.23	(C) 2.43	(D) 18.61
#8:	(C) 80.33	(A) 0.21	(B) 6.72	(D) 0.48	(E) 0.19
#9:	(C) 94.30	(A) 1.14	(B) 1.41	(D) 0.55	(E) 0.15
#10:	(E) 53.90	(A) 2.15	(B) 7.76	(C) 0.48	(D) 5.08
#11:	(E) 80.57	(A) 0.88	(B) 0.19	(C) 0.22	(D) 1.52
#12:	(E) 95.30	(A) 0.02	(B) 0.03	(C) 0.81	(D) 0.29
#13:	(C) 46.72	(A) 1.17	(B) 2.23	(D) 2.20	(E) 0.61
#14:	(E) 27.41	(A) 4.29	(B) 0.96	(C) 1.19	(D) 3.21
#15:	(B) 57.38	(A) 10.07	(C) 0.80	(D) 0.54	(E) 0.33
#16:	(D) 29.38	(A) 10.99	(B) 7.41	(C) 1.82	(E) 0.63
#17:	(D) 16.19	(A) 0.56	(B) 6.69	(C) 0.91	(E) 0.54
#18:	(A) 50.03	(B) 3.02	(C) 3.33	(D) 3.42	(E) 3.08
#19:	(C) 10.36	(A) 1.03	(B) 0.93	(D) 1.10	(E) 0.90
#20:	(B) 24.06	(A) 0.57	(C) 0.66	(D) 0.72	(E) 0.49
#21:	(D) 13.96	(A) 1.04	(B) 4.16	(C) 1.96	(E) 0.66
#22:	(C) 8.49	(A) 3.29	(B) 1.32	(D) 2.21	(E) 0.70
#23:	(B) 4.53	(A) 1.23	(C) 1.80	(D) 2.25	(E) 3.38
#24:	(D) 2.13	(A) 2.17	(B) 0.35	(C) 0.38	(E) 0.36
#25:	(E) 4.88	(A) 0.47	(B) 0.45	(C) 0.88	(D) 1.72

1st AMC 10, 2000

1. In the year 2001, the United States will host the International Mathematical Olympiad. Let $I, M,$ and O be distinct positive integers such that the product $I \cdot M \cdot O = 2001$. What is the largest possible value of the sum $I + M + O$?

   (A) 23   (B) 55   (C) 99   (D) 111   (E) 671

2. $2000(2000^{2000}) =$

   (A) 2000^{2001} (B) 4000^{2000} (C) 2000^{4000}
   (D) $4,000,000^{2000}$ (E) $2000^{4,000,000}$

3. Each day, Jenny ate 20% of the jellybeans that were in her jar at the beginning of that day. At the end of the second day, 32 remained. How many jellybeans were in the jar originally?

   (A) 40   (B) 50   (C) 55   (D) 60   (E) 75

4. Chandra pays an on-line service provider a fixed monthly fee plus an hourly charge for connect time. Her December bill was $12.48, but in January her bill was $17.54 because she used twice as much connect time as in December. What is the fixed monthly fee?

   (A) $2.53   (B) $5.06   (C) $6.24
   (D) $7.42   (E) $8.77

5. Points M and N are the midpoints of sides PA and PB of $\triangle PAB$. As P moves along a line that is parallel to side AB, how many of the four quantities listed below change?

the length of the segment MN
the perimeter of $\triangle PAB$
the area of $\triangle PAB$
the area of trapezoid $ABNM$

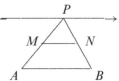

(A) 0   (B) 1   (C) 2   (D) 3   (E) 4

6. The Fibonacci sequence $1, 1, 2, 3, 5, 8, 13, 21, \ldots$ starts with two 1s, and each term afterwards is the sum of its two predecessors. Which one of the ten digits is the last to appear in the units position of a number in the Fibonacci sequence?

(A) 0   (B) 4   (C) 6   (D) 7   (E) 9

7. In rectangle $ABCD$, $AD = 1$, P is on $\overline{AB}$, and $\overline{DB}$ and $\overline{DP}$ trisect $\angle ADC$. What is the perimeter of $\triangle BDP$?

(A) $3 + \dfrac{\sqrt{3}}{3}$ (B) $2 + \dfrac{4\sqrt{3}}{3}$

(C) $2 + 2\sqrt{2}$ (D) $\dfrac{3 + 3\sqrt{5}}{2}$

(E) $2 + \dfrac{5\sqrt{3}}{3}$

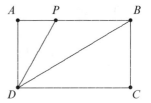

8. At Olympic High School, 2/5 of the freshmen and 4/5 of the sophomores took the AMC 10. Given that the number of freshmen and sophomore contestants was the same, which of the following must be true?

(A) There are five times as many sophomores as freshmen.

(B) There are twice as many sophomores as freshmen.

(C) There are as many freshmen as sophomores.

(D) There are twice as many freshmen as sophomores.

(E) There are five times as many freshmen as sophomores.

9. If $|x - 2| = p$, where $x < 2$, then $x - p =$

   (A) -2 (B) 2 (C) $2 - 2p$ (D) $2p - 2$ (E) $|2p - 2|$

10. The sides of a triangle with positive area have lengths 4, 6, and x. The sides of a second triangle with positive area have lengths 4, 6, and y. What is the smallest positive number that is **not** a possible value of $|x - y|$?

    (A) 2 (B) 4 (C) 6 (D) 8 (E) 10

11. Two different prime numbers between 4 and 18 are chosen. When their sum is subtracted from their product, which of the following numbers could be obtained?

    (A) 21 (B) 60 (C) 119 (D) 180 (E) 231

12. Figures 0, 1, 2, and 3 consist of 1, 5, 13, and 25 nonoverlapping unit squares, respectively. If the pattern were continued, how many nonoverlapping unit squares would there be in figure 100?

    (A) 10401 (B) 19801 (C) 20201 (D) 39801 (E) 40801

    figure 0   figure 1   figure 2   figure 3

13. There are 5 yellow pegs, 4 red pegs, 3 green pegs, 2 blue pegs, and 1 orange peg to be placed on a triangular peg board. In how many ways can the pegs be placed so that no (horizontal) row or (vertical) column contains two pegs of the same color?

    (A) 0

    (B) 1

    (C) $5! \cdot 4! \cdot 3! \cdot 2! \cdot 1!$

    (D) $15!/(5! \cdot 4! \cdot 3! \cdot 2! \cdot 1!)$

    (E) $15!$

    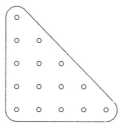

14. Mrs. Walter gave an exam in a mathematics class of five students. She entered the scores in random order into a spreadsheet, which recalculated the class average after each score was entered. Mrs. Walter noticed that after each score was entered, the average was always an integer. The scores (listed in ascending order) were 71, 76, 80, 82, and 91. What was the last score Mrs. Walter entered?

   (A) 71 (B) 76 (C) 80 (D) 82 (E) 91

15. Two non-zero real numbers, a and b, satisfy $ab = a - b$. Which of the following is a possible value of $a/b + b/a - ab$?

   (A) -2 (B) $-\dfrac{1}{2}$ (C) $\dfrac{1}{3}$ (D) $\dfrac{1}{2}$ (E) 2

16. The diagram shows 28 lattice points, each one unit from its nearest neighbors. Segment AB meets segment CD at E. Find the length of segment AE.

   (A) $4\sqrt{5}/3$
   (B) $5\sqrt{5}/3$
   (C) $12\sqrt{5}/7$
   (D) $2\sqrt{5}$
   (E) $5\sqrt{65}/9$

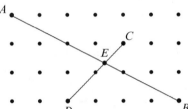

17. Boris has an incredible coin changing machine. When he puts in a quarter, it returns five nickels; when he puts in a nickel, it returns five pennies; and when he puts in a penny, it returns five quarters. Boris starts with just one penny. Which of the following amounts could Boris have after using the machine repeatedly?

   (A) $3.63 (B) $5.13 (C) $6.30
   (D) $7.45 (E) $9.07

18. Charlyn walks completely around the boundary of a square whose sides are each 5 km long. From any point on her path she can see exactly 1 km horizontally in all directions. What is the area of the region consisting of all points Charlyn can see during her walk, expressed in square kilometers and rounded to the nearest whole number?

   (A) 24 (B) 27 (C) 39 (D) 40 (E) 42

19. Through a point on the hypotenuse of a right triangle, lines are drawn parallel to the legs of the triangle so that the triangle is divided into a square and two smaller right triangles. The area of one of the two small right triangles is m times the area of the square. The ratio of the area of the other small right triangle to the area of the square is

    (A) $\dfrac{1}{2m+1}$ (B) m (C) $1-m$ (D) $\dfrac{1}{4m}$ (E) $\dfrac{1}{8m^2}$

20. Let A, M, and C be nonnegative integers such that $A+M+C=10$. What is the maximum value of $A \cdot M \cdot C + A \cdot M + M \cdot C + C \cdot A$?

    (A) 49 (B) 59 (C) 69 (D) 79 (E) 89

21. If all alligators are ferocious creatures and some creepy crawlers are alligators, which statement(s) **must** be true?

    I. All alligators are creepy crawlers.

    II. Some ferocious creatures are creepy crawlers.

    III. Some alligators are not creepy crawlers.

    (A) I only (B) II only (C) III only
    (D) II and III only (E) None must be true

22. One morning each member of Angela's family drank an 8-ounce mixture of coffee with milk. The amounts of coffee and milk varied from cup to cup, but were never zero. Angela drank a quarter of the total amount of milk and a sixth of the total amount of coffee. How many people are in the family?

    (A) 3 (B) 4 (C) 5 (D) 6 (E) 7

23. When the mean, median, and mode of the list

    $$10, 2, 5, 2, 4, 2, x$$

    are arranged in increasing order, they form a non-constant arithmetic progression. What is the sum of all possible real values of x?

    (A) 3 (B) 6 (C) 9 (D) 17 (E) 20

24. Let f be a function for which $f(x/3) = x^2 + x + 1$. Find the sum of all values of z for which $f(3z) = 7$.

    (A) $-1/3$ (B) $-1/9$ (C) 0 (D) $5/9$ (E) $5/3$

25. In year N, the 300th day of the year is a Tuesday. In year $N+1$, the 200th day is also a Tuesday. On what day of the week did the 100th day of year $N-1$ occur?

    (A) Thursday   (B) Friday   (C) Saturday
    (D) Sunday   (E) Monday

The 2001 AMC 10, AMC 12 were distributed to 5162 schools. The AMC 10 was written by 106,994 students. The number of qualifiers through the AMC 10 was 1201. The following table lists the percent of AIME qualifiers on the AMC 10 who gave each answer to each question. The correct answer is the one in the first column. Problem 13 presented a special challenge to the committee. Several students appealed their answer of C, arguing that if the n pegs of a given color are distinguishable, then there are $n!$ ways to place them in their positions, so that answer C would be correct. The committee decided to award credit for the answer C.

```
ANSWER
   #1:  (E) 82.01  (A) 0.08   (B) 14.53  (C) 0.17   (D) 0.59
   #2:  (A) 98.99  (B) 0.17   (C) 0.00   (D) 0.59   (E) 0.17
   #3:  (B) 98.40  (A) 1.44   (C) 0.17   (D) 0.00   (E) 0.00
   #4:  (D) 99.16  (A) 0.00   (B) 0.84   (C) 0.00   (E) 0.00
   #5:  (B) 59.97  (A) 6.33   (C) 8.28   (D) 6.93   (E) 1.10
   #6:  (C) 82.01  (A) 6.25   (B) 0.42   (D) 1.52   (E) 1.10
   #7:  (B) 77.28  (A) 0.76   (C) 1.18   (D) 0.84   (E) 1.60
   #8:  (D) 94.76  (A) 0.00   (B) 3.21   (C) 2.03   (E) 0.00
   #9:  (C) 83.02  (A) 2.11   (B) 4.81   (D) 2.03   (E) 1.44
  #10:  (D) 78.38  (A) 1.27   (B) 1.10   (C) 1.27   (E) 6.33
  #11:  (C) 98.31  (A) 0.42   (B) 0.00   (D) 0.00   (E) 0.08
  #12:  (C) 78.46  (A) 0.42   (B) 4.48   (D) 0.51   (E) 0.68
  #13:  (B) 80.24  (A) 1.18   (C) 8.53   (D) 1.44   (E) 0.08
  #14:  (C) 90.29  (A) 1.86   (B) 2.11   (D) 0.59   (E) 0.08
  #15:  (E) 76.18  (A) 1.01   (B) 0.34   (C) 0.25   (D) 1.86
  #16:  (B) 52.62  (A) 1.94   (C) 1.94   (D) 1.35   (E) 2.03
  #17:  (D) 74.07  (A) 0.34   (B) 0.08   (C) 1.52   (E) 0.25
  #18:  (C) 65.29  (A) 2.03   (B) 1.60   (D) 12.58  (E) 2.03
  #19:  (D) 18.67  (A) 2.11   (B) 13.94  (C) 4.90   (E) 1.52
  #20:  (C) 87.92  (A) 1.35   (B) 0.25   (D) 0.76   (E) 0.08
  #21:  (B) 75.25  (A) 1.35   (C) 0.59   (D) 12.58  (E) 3.46
  #22:  (C) 37.08  (A) 0.84   (B) 1.86   (D) 1.77   (E) 0.25
  #23:  (E) 15.37  (A) 5.74   (B) 1.52   (C) 1.35   (D) 2.96
  #24:  (B) 31.84  (A) 10.22  (C) 1.01   (D) 1.10   (E) 0.42
  #25:  (A) 46.11  (B) 4.39   (C) 3.97   (D) 4.56   (E) 4.81
```

50th Anniversary AHSME

1950-10 After rationalizing the numerator of $\dfrac{\sqrt{3}-\sqrt{2}}{\sqrt{3}}$, the denominator in simplest form is

(A) $\sqrt{3}(\sqrt{3}+\sqrt{2})$ (B) $\sqrt{3}(\sqrt{3}-\sqrt{2})$ (C) $3-\sqrt{3}\sqrt{2}$
(D) $3+\sqrt{6}$ (E) none of these answers

1951-48 The area of a square inscribed in a semicircle is to the area of the square inscribed in the entire circle as:

(A) $1:2$ (B) $2:3$ (C) $2:5$ (D) $3:4$ (E) $3:5$

1952-44 If an integer of two digits is k times the sum of its digits, the number formed by interchanging the digits is the sum of the digits multiplied by

(A) $9-k$ (B) $10-k$ (C) $11-k$
(D) $k-1$ (E) $k+1$

1953-50 One of the sides of a triangle is divided into segments of 6 and 8 units by the point of tangency of the inscribed circle. If the radius of the circle is 4, then the length of the shortest side is

(A) 12 units (B) 13 units (C) 14 units
(D) 15 units (E) 16 units

1954-38 If $\log 2 = .3010$ and $\log 3 = .4771$, the value of x when $3^{x+3} = 135$ is approximately

(A) 5 (B) 1.47 (C) 1.67 (D) 1.78 (E) 1.63

1955-33 Henry starts a trip when the hands of the clock are together between 8 AM and 9 AM he arrives at his destination between 2 PM and 3 PM when the hands are exactly 180° apart. The trip takes

(A) 6 hr. (B) 6 hr. $43\frac{7}{11}$ min. (C) 5 hr. $16\frac{4}{11}$ min.

(D) 6 hr. 30 min. (E) none of these

1956-39 The hypotenuse c and one side a of a right triangle are consecutive integers. The square of the second side is

(A) ca (B) $\dfrac{c}{a}$ (C) $c + a$ (D) $c - a$ (E) none of these

1957-26 From a point P within a triangle, line segments are drawn to the vertices. A necessary and sufficient condition that the three triangles formed have equal areas is that the point P be

(A) the center of the inscribed circle.

(B) the center of the circumscribed circle.

(C) such that the three angles formed at P each be 120°.

(D) the intersection of the altitudes of the triangle.

(E) the intersection of the medians of the triangle.

1958-45 A check is written for x dollars and y cents, both x and y two-digit numbers. In error it is cashed for y dollars and x cents, the incorrect amount exceeding the correct amount by $17.82. Then

(A) x cannot exceed 70

(B) y can equal $2x$

(C) the amount of the check cannot be a multiple of 5

(D) the incorrect amount can be twice the correct amount

(E) the sum of the digits of the correct amount is divisible by 9

1959-22 The line joining the midpoints of the diagonals of a trapezoid has length 3. If the longer base is 97, then the shorter base is

(A) 94 (B) 92 (C) 91 (D) 90 (E) 89

1960-19 Consider equation **I** : $x + y + z = 46$ where $x, y,$ and z are positive integers, and the equation **II** : $x + y + z + w = 46$ where $x, y, z,$ and w are positive integers. Then

(A) **I** can be solved in consecutive integers.

(B) **I** can be solved in consecutive even integers.

(C) **II** can be solved in consecutive integers.

(D) **II** can be solved in consecutive even integers.

(E) **II** can be solved in consecutive odd integers.

1961-5 Let $S = (x-1)^4 + 4(x-1)^3 + 6(x-1)^2 + 4(x-1) + 1$. Then $S =$

(A) $(x-2)^4$ (B) $(x-1)^4$ (C) x^4
(D) $(x+1)^4$ (E) $x^4 + 1$

1962-27 Let $a \, \text{\textcircled{L}} \, b$ represent the operation on two numbers, a and b, which selects the larger of the two numbers, with $a \, \text{\textcircled{L}} \, a = a$. Let $a \, \text{\textcircled{S}} \, b$ represent the operation on two numbers, a and b, which selects the smaller of the two numbers, with $a \, \text{\textcircled{S}} \, a = a$. Which of the following rules is (are) correct?

(1) $a \, \text{\textcircled{L}} \, b = b \, \text{\textcircled{L}} \, a$,
(2) $a \, \text{\textcircled{L}} \, (b \, \text{\textcircled{L}} \, c) = (a \, \text{\textcircled{L}} \, b) \, \text{\textcircled{L}} \, c$,
(3) $a \, \text{\textcircled{S}} \, (b \, \text{\textcircled{L}} \, c) = (a \, \text{\textcircled{S}} \, b) \, \text{\textcircled{L}} \, (a \, \text{\textcircled{S}} \, c)$.

(A) (1) only   (B) (2) only   (C) (1) and (2) only
(D) (1) and (2) only   (E) all three

1963-37 Given points $P_1, P_2, \ldots, P_7$ on a straight line, in the order stated (not necessarily evenly spaced). Let P be an arbitrary point selected on the line and let s be the sum of the undirected lengths

$$PP_1, PP_2, \ldots, PP_7.$$

Then s is smallest if and only if the point P is

(A) midway between P_1 and P_7

(B) midway between P_2 and P_6

(C) midway between P_3 and P_5

(D) at P_4 (E) at P_1

1964-15 A line through the point $(-a, 0)$ cuts from the second quadrant a triangular region with area T. The equation for the line is

(A) $2Tx + a^2y + 2aT = 0$ (B) $2Tx - a^2y + 2aT = 0$
(C) $2Tx + a^2y - 2aT = 0$ (D) $2Tx - a^2y - 2aT = 0$
(E) none of these

1965-29 Of 28 students taking at least one subject, the number taking Mathematics and English only equals the number taking Mathematics only. No student takes English only or History only, and six students take Mathematics and History, but no English. The number taking English and History only is five times the number taking all three subjects. If the number taking all three subjects is even and non-zero, the number taking English and Mathematics only is

(A) 5 (B) 6 (C) 7 (D) 8 (E) 9

1966-39 In base b the expanded fraction F_1 becomes $.3737\ldots = .\overline{37}$, and the expanded fraction F_2 becomes $.7373\ldots = .\overline{73}$. In base a the expanded fraction F_1 becomes $.2525\ldots = .\overline{25}$, and the expanded fraction F_2 becomes $.5252\ldots = .\overline{52}$. The sum of a and b, each written in base ten, is

(A) 24 (B) 22 (C) 21 (D) 20 (E) 19

1967-31 Let $D = a^2 + b^2 + c^2$, where a and b are consecutive integers and $c = ab$. Then $\sqrt{D}$ is

(A) always an even integer

(B) sometimes an odd integer, sometimes not

(C) always an odd integer

(D) sometimes rational, sometimes not

(E) always irrational

1968-32 A and B move uniformly along two straight paths intersecting at right angles in point O. When A is at O, B is 500 yards from O. In two minutes they are equidistant from O, and in eight minutes more they are again equidistant from O. Then the ratio of A's speed to B's speed is

(A) $4:5$ (B) $5:6$ (C) $2:3$ (D) $5:8$ (E) $1:2$

1969-29 If $x = t^{1/(t-1)}$ and $y = t^{t/(t-1)}, t > 0, t \neq 1$, a relation between x and y is

(A) $y^x = x^{1/y}$ (B) $y^{1/x} = x^y$ (C) $x^y = y^x$
(D) $x^x = y^y$ (E) none of these

1970-25 For every real number x, let $\lfloor x \rfloor$ be the greatest integer which is less than or equal to x. If the postal rate for first class mail is six cents for every ounce or portion thereof, then the cost in cents of first-class postage on a letter weighing W ounces is always

(A) $6W$ (B) $6\lfloor W \rfloor$ (C) $6(\lfloor W \rfloor - 1)$
(D) $6(\lfloor W \rfloor + 1)$ (E) $-6\lfloor -W \rfloor$

1971-31 Quadrilateral $ABCD$ is inscribed in a circle with side AD, a diameter of length 4. If sides AB and BC each have length 1, then CD has length

(A) $7/2$
(B) $5\sqrt{2}/2$
(C) $\sqrt{11}$
(D) $\sqrt{13}$
(E) $2\sqrt{3}$

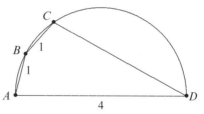

1972-35 Equilateral triangle ABP with side AB of length two inches is placed inside a square $AXYZ$ with side of length four inches so that B is on side AX. The triangle is rotated clockwise about B, then P, and so on along the sides of the square until P, A, and B all return to their original positions. The length of the path in inches traversed by vertex P is equal to

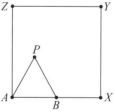

(A) $20\pi/3$ (B) $32\pi/3$ (C) 12π (D) $40\pi/3$ (E) 15π

1973-31 In the following equation, each letter represents uniquely a different digit in base ten:
$$(YE) \cdot (ME) = TTT$$
The sum $E + M + T + Y$ equals

(A) 19 (B) 20 (C) 21 (D) 22 (E) 24

1974-20 Let $T = \dfrac{1}{3-\sqrt{8}} - \dfrac{1}{\sqrt{8}-\sqrt{7}} + \dfrac{1}{\sqrt{7}-\sqrt{6}} - \dfrac{1}{\sqrt{6}-\sqrt{5}} + \dfrac{1}{\sqrt{5}-2}$; then

(A) $T < 1$ (B) $T = 1$ (C) $1 < T < 2$ (D) $T > 2$

(E) $T = \dfrac{1}{(3-\sqrt{8})(\sqrt{8}-\sqrt{7})(\sqrt{7}-\sqrt{6})(\sqrt{6}-\sqrt{5})(\sqrt{5}-2)}$

1975-25 A woman, her brother, her son, and her daughter are chess players (all relations by birth). The worst player's twin (who is one of the four players) and the best player are of opposite sex. The worst player and the best player are the same age. Who is the worst player?

(A) the woman (B) her son (C) her brother (D) her daughter
(E) No solution is consistent with the given information

1976-30 How many distinct ordered triples (x, y, z) satisfy the equations
$$x + 2y + 4z = 12$$
$$xy + 4yz + 2xz = 22$$
$$xyz = 6$$

(A) none (B) 1 (C) 2 (D) 4 (E) 6

1977-8 For every triple (a, b, c) of non-zero real numbers, form the number
$$\dfrac{a}{|a|} + \dfrac{b}{|b|} + \dfrac{c}{|c|} + \dfrac{abc}{|abc|}.$$
The set of all numbers formed is

(A) $\{0\}$ (B) $\{-4, 0, 4\}$ (C) $\{-4, -2, 0, 2, 4\}$
(D) $\{-4, -2, 2, 4\}$ (E) none of the these

1978-22 The following four statements, and only these are found on a card:

> On this card exactly one statement is false.
> On this card exactly two statements are false.
> On this card exactly three statements are false.
> On this card exactly four statements are false.

(Assume each statement is either true or false.) Among them the number of false statements is exactly

(A) 0 (B) 1 (C) 2 (D) 3 (E) 4

1979-26 The function f satisfies the functional equation
$$f(x) + f(y) = f(x+y) - xy - 1$$
for every pair x, y of real numbers. If $f(1) = 1$, then the number of integers $n \neq 1$ for which $f(n) = n$ is

(A) 0 (B) 1 (C) 2 (D) 3 (E) infinite

1980-22 For each real number x, let $f(x)$ be the minimum of the numbers $4x + 1, x + 2,$ and $-2x + 4$. Then the maximum value of $f(x)$ is

(A) $\frac{1}{3}$ (B) $\frac{1}{2}$ (C) $\frac{2}{3}$ (D) $\frac{5}{2}$ (E) $\frac{8}{3}$

1981-24 If θ is a constant such that $0 < \theta < \pi$ and $z + 1/z = 2\cos\theta$, then for each positive integer n, $z^n + /1/z^n$ equals

(A) $2\cos\theta$ (B) $2^n \cos\theta$ (C) $2\cos^n \theta$ (D) $2\cos n\theta$ (E) $2^n \cos^n \theta$

1982-16 In the figure below, a wooden cube has edges of length three meters. Square holes of side one meter, centered in each face, are cut through to the opposite face. The edges of the whole are parallel to the edges of the cube. The entire surface area including the inside, in square meters, is

(A) 54 (B) 72 (C) 76 (D) 84 (E) 86

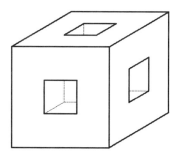

1983-26 The probability that event A occurs is 3/4; the probability that event B occurs is 2/3. Let p be the probability that both A and B occur. The smallest interval necessarily containing p is the interval

(A) $\left[\frac{1}{12}, \frac{1}{2}\right]$ (B) $\left[\frac{5}{12}, \frac{1}{2}\right]$ (C) $\left[\frac{1}{2}, \frac{2}{3}\right]$ (D) $\left[\frac{5}{12}, \frac{2}{3}\right]$ (E) $\left[\frac{1}{12}, \frac{2}{3}\right]$

1984-11 A calculator has a key which replaces the displayed entry with its square, and another key which replaces the displayed entry with its reciprocal. Let y be the final result if one starts with an entry $x \neq 0$ and alternately squares and reciprocates n times each. Assuming the calculator is completely accurate (e.g., no roundoff or overflow), then y equals

(A) $x^{((-2)^n)}$ (B) x^{2n} (C) x^{-2n}
(D) $x^{-(2^n)}$ (E) $x^{((-1)^n 2n)}$

1985-24 A non-zero digit is chosen in such a way that the probability of choosing digit d is $\log_{10}(d+1) - \log_{10} d$. The probability that the digit 2 is chosen is exactly $1/2$ the probability that the digit chosen is in the set

(A) $\{2, 3\}$ (B) $\{3, 4\}$ (C) $\{4, 5, 6, 7, 8\}$
(D) $\{5, 6, 7, 8, 9\}$ (E) $\{4, 5, 6, 7, 8, 9\}$

1986-14 Suppose hops, skips and jumps are specific units of length. If b hops equals c skips, d jumps equals e hops, and f jumps equals g meters, then one meter equals how many skips?

(A) $\dfrac{bdg}{cef}$ (B) $\dfrac{cdf}{beg}$ (C) $\dfrac{cdg}{bef}$ (D) $\dfrac{cef}{bdg}$ (E) $\dfrac{ceg}{bdf}$

1987-12 In an office, at various times during the day the boss gives the secretary a letter to type, each time putting the letter on top of the pile in the secretary's in-box. When there is time, the secretary takes the top letter off the pile and types it. If there are five letters in all, and the boss delivers them in the order 1 2 3 4 5, which of the following could *not* be the order in which the secretary types them?

(A) 12345   (B) 24351   (C) 32415
(D) 45231   (E) 54321

1988-6 A figure is an equiangular parallelogram if and only if it is a

(A) rectangle   (B) regular polygon   (C) rhombus
(D) square   (E) trapezoid

1989-23 A particle moves through the first quadrant as follows. During the first minute it moves from the origin to $(1,0)$. Thereafter, it continues to follow the directions indicated in the figure, going back and forth between the positive x and y axes, moving one unit of distance parallel to an axis in each minute. At which point will the particle be after exactly 1989 minutes?

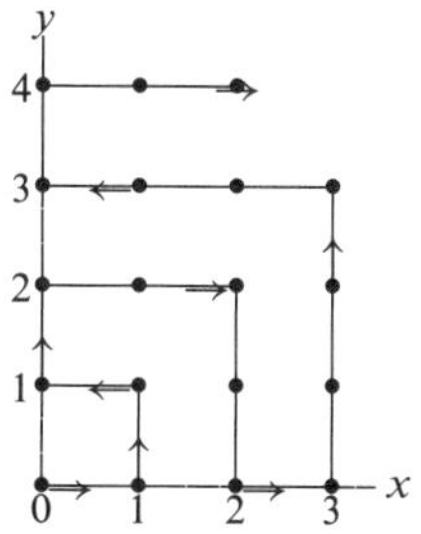

(A) $(35, 44)$

(B) $(36, 45)$

(C) $(37, 45)$

(D) $(44, 35)$

(E) $(45, 36)$

1990-14 An acute isosceles triangle, ABC, is inscribed in a circle. Through B and C, tangents to the circle are drawn, meeting at point D. If $\angle ABC = \angle ACB = 2\angle D$ and x is the radian measure of $\angle A$, then $x=$

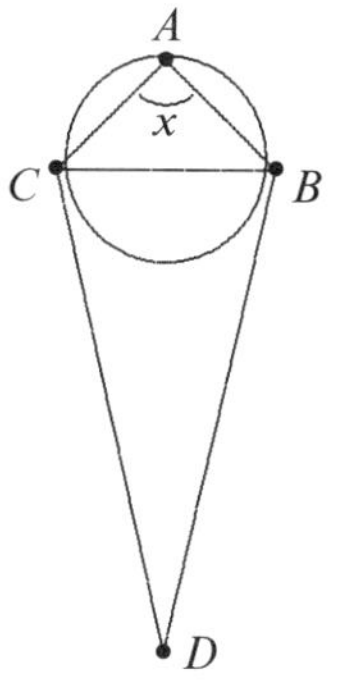

(A) $\dfrac{3}{7}\pi$

(B) $\dfrac{4}{9}\pi$

(C) $\dfrac{5}{11}\pi$

(D) $\dfrac{6}{13}\pi$

(E) $\dfrac{7}{15}\pi$

1991-28 Initially an urn contains 100 black marbles and 100 white marbles. Repeatedly, three marbles are removed from the urn and replaced from a pile outside the urn as follows:

MARBLES REMOVED	REPLACED WITH
3 black	1 black
2 black, 1 white	1 black, 1 white
1 black, 2 white	2 white
3 white	1 black, 1 white.

Which of the following sets of marbles could be the contents of the urn after repeated applications of this procedure?

(A) 2 black marbles (B) 2 white marbles
(C) 1 black marble (D) 1 black and 1 white marble
(E) 1 white marble

1992-14 Which of the following equations have the same graph?
I. $y = x - 2$ II. $y = \dfrac{x^2 - 4}{x + 2}$ III. $(x + 2)y = x^2 - 4$

(A) I and II only (B) I and III only
(C) II and III only (D) I, II and III
(E) None. All the equations have different graphs

1993-22 Twenty cubical blocks are arranged as shown. First, 10 are arranged in a triangular pattern; then a layer of 6, arranged in a triangular pattern, is centered on the 10; then a layer of 3, arranged in a triangular pattern, is centered on the 6; and finally one block is centered on top of the third layer. The blocks in the bottom layer are numbered 1 through 10 in some order. Each block in layers 2, 3 and 4 is assigned the number which is the sum of the numbers assigned to the three blocks on which it rests. Find the smallest possible number which could be assigned to the top block.

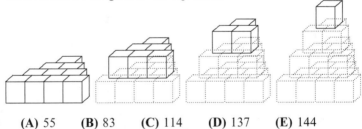

(A) 55 (B) 83 (C) 114 (D) 137 (E) 144

1994-6 In the sequence
$$\ldots, a, b, c, d, 0, 1, 1, 2, 3, 5, 8, \ldots$$
each term is the sum of the two terms to its left. Find a.

(A) -3 (B) -1 (C) 0 (D) 1 (E) 3

1995-30 A large cube is formed by stacking 27 unit cubes. A plane is perpendicular to one of the internal diagonals of the large cube and bisects that diagonal. The number of unit cubes that the plane intersects is

(A) 16   (B) 17   (C) 18   (D) 19   (E) 20

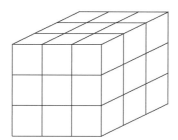

1996-27 Consider two solid spherical balls, one centered at $(0, 0, \frac{21}{2})$ with radius 6, and the other centered at $(0, 0, 1)$ with radius $\frac{9}{2}$. How many points (x, y, z) with only integer coordinates (lattice points) are there in the intersection of the balls?

(A) 7   (B) 9   (C) 11   (D) 13   (E) 15

1997-29 Call a positive real number *special* if it has a decimal representation that consists entirely of digits 0 and 7. For example, $\frac{700}{99} = 7.\overline{07} = 7.070707\ldots$ and 77.007 are special numbers. What is the smallest n such that 1 can be written as a sum of n special numbers?

(A) 7   (B) 8   (C) 9   (D) 10

(E) The number 1 cannot be represented as a sum of finitely many special numbers.

1998-22 What is the value of the expression

$$\frac{1}{\log_2 100!} + \frac{1}{\log_3 100!} + \frac{1}{\log_4 100!} + \cdots + \frac{1}{\log_{100} 100!}?$$

(A) 0.01   (B) 0.1   (C) 1   (D) 2   (E) 10

1999-18 How many zeros does $f(x) = \cos(\log(x))$ have on the interval $0 < x < 1$?

(A) 0   (B) 1   (C) 2   (D) 10   (E) infinitely many

Answers:

1950 ... D	1951 ... C	1952 ... C	1953 ... B	1954 ... B
1955 ... A	1956 ... C	1957 ... E	1958 ... B	1959 ... C
1960 ... C	1961 ... C	1962 ... E	1963 ... D	1964 ... B
1965 ... A	1966 ... E	1967 ... C	1968 ... C	1969 ... C
1970 ... E	1971 ... A	1972 ... D	1973 ... C	1974 ... D
1975 ... B	1976 ... E	1977 ... B	1978 ... D	1979 ... B
1980 ... E	1981 ... D	1982 ... B	1983 ... D	1984 ... A
1985 ... C	1986 ... D	1987 ... D	1988 ... A	1989 ... D
1990 ... A	1991 ... B	1992 ... E	1993 ... C	1994 ... A
1995 ... D	1996 ... D	1997 ... B	1998 ... C	1999 ... E

46th AHSME solutions, 1995

1. **(B)** The average changes from $(87 + 83 + 88)/3 = 86$ to $(87 + 83 + 88 + 90)/4 = 87$, an increase of 1.

2. **(D)** Square both sides of the given equation to obtain $2 + \sqrt{x} = 9$. Thus $\sqrt{x} = 7$, and $x = 49$, which satisfies the given equation.

3. **(B)** The total price advertised on television is
$$\$29.98 + \$29.98 + \$29.98 + \$9.98 = \$99.92,$$
so this is $\$99.99 - \$99.92 = \$0.07$ less than the in-store price.

OR

The three payments are each 2 cents less than $30, and the shipping & handling charge is 2 cents less than $10, so the total price advertised on television is 8 cents less than $100. The total in-store price is 1 cent less than $100, so the amount saved by buying the appliance from the television advertiser is 7 cents.

4. **(B)** Since $M = 0.3Q = 0.3(0.2P) = 0.06P$ and $N = 0.5P$, we have
$$\frac{M}{N} = \frac{0.06P}{0.5P} = \frac{6}{50} = \frac{3}{25}.$$

5. **(C)** The number of ants is approximately the product
$$(300 \text{ ft}) \times (400 \text{ ft}) \times (12 \text{ in/ft})^2 \times (3 \text{ ants/in}^2) = 300 \times 400 \times 144 \times 3$$
ants, which is $3 \times 4 \times 1.44 \times 3 \times 10^{2+2+2} \approx 50 \times 10^6$.

6. **(C)** Think of A as the bottom. Fold B up to be the back. Then x folds upward to become the left side and C folds forward to become the right side, so C is opposite x.

7. **(C)** The length of the flight path is approximately the circumference of Earth at the equator, which is

$$C = 2\pi \cdot 4000 = 8000\pi \text{ miles.}$$

The time required is

$$\frac{8000\pi}{500} = 16\pi \begin{cases} > 16(3.1) = 49.6 \text{ hours} \\ < 16(3.2) = 51.2 \text{ hours,} \end{cases}$$

so the best choice is 50 hours. **Query.** What is a negligible height; i.e., for which heights above the equator would the flight-time be closer to choice **(C)** than to **(D)**?

8. **(C)** Because $\triangle ABC$ is a right triangle, the *Pythagorean Theorem* implies that $BA = 10$. Since $\triangle DBE \sim \triangle ABC$,

$$\frac{BD}{BA} = \frac{DE}{AC}. \quad \text{So} \quad BD = \frac{DE}{AC}(BA) = \frac{4}{6}(10) = \frac{20}{3}.$$

OR

Since $\sin B = DE/BD$, we have $BD = DE/\sin B$. Moreover, $BA = 10$ by the *Pythagorean Theorem*, so $\sin B = AC/BA = 3/5$. Hence $BD = 4 \div 3/5 = 20/3$.

9. **(D)** Since all the acute angles in the figure measure $45°$, all the triangles must be isosceles right triangles. It follows that all the triangles must enclose one, two or four of the eight small triangular regions. Besides the eight small triangles, there are four triangles that enclose two of the small triangular regions and four triangles that enclose four, making a total of 16.

10. **(E)** Let O be the origin, and let A and B denote the points where $y = 6$ intersects $y = x$ and $y = -x$ respectively. Let $\overline{OL}$ denote the altitude to side $\overline{AB}$ of $\triangle OAB$. Then $OL = 6$. Also, $AL = BL = 6$. Thus, the area of $\triangle OAB$ is

$$\frac{1}{2}(AB)(OL) = \frac{1}{2} \cdot 12 \cdot 6 = 36.$$

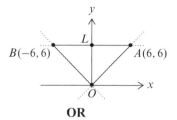

OR

Let $A' = (6, 0)$. Then $\triangle A'OA \cong \triangle LOB$, so the area of triangle AOB equals the area of square $A'OLA$, which is $6^2 = 36$.

OR

Use the determinant formula for the area of the triangle:

$$\frac{1}{2} \begin{vmatrix} 0 & 0 & 1 \\ 6 & 6 & 1 \\ -6 & 6 & 1 \end{vmatrix} = 36.$$

11. **(C)** Condition (*i*) requires that a be one of the two digits, 4 or 5. Condition (*ii*) requires that d be one of the two digits, 0 or 5. Condition (*iii*) requires that the ordered pair (b, c) be one of these six ordered pairs:

$$(3, 4), \ (3, 5), \ (3, 6), \ (4, 5), \ (4, 6), \ (5, 6).$$

Therefore, there are $2 \times 2 \times 6 = 24$ numbers N satisfying the conditions.

12. **(D)** Since f is a linear function, it has the form $f(x) = mx + b$. Because $f(1) \leq f(2)$, we have $m \geq 0$. Similarly, $f(3) \geq f(4)$ implies $m \leq 0$. Hence, $m = 0$, and f is a constant function. Thus, $f(0) = f(5) = 5$.

13. **(C)** The addition in the columns containing the ten-thousands and hundred-thousands digits is incorrect. The only digit common to both these columns is 2. Changing these 2's to 6's makes the arithmetic correct. Changing the other two 2's to 6's has no effect on the correctness of the remainder of the addition, and no digit other than 2 could be changed to make the addition correct. Thus, $d = 2$, $e = 6$, and $d + e = 8$.

14. **(E)** Since $f(3) = a(3)^4 - b(3)^2 + 3 + 5$ and $f(-3) = a(-3)^4 - b(-3)^2 - 3 + 5$, it follows that $f(3) - f(-3) = 6$. Thus, $f(3) = f(-3) + 6 = 2 + 6 = 8$.

Note. For any x, $f(x) - f(-x) = 2x$, so $f(x) = f(-x) + 2x$.

OR

Since $2 = f(-3) = 81a - 9b - 3 + 5$ we have $b = 9a$. Thus $f(3) = 81a - 9b + 3 + 5 = 81a - 9(9a) + 8 = 8$.

15. **(D)** With the first jump, the bug moves to point 1, with the second to 2, with the third to 4 and with the fourth it returns to 1. Thereafter, every third jump it returns to 1. Thus, after $n > 0$ jumps, the bug will be on 1, 2 or 4, depending on whether n is of the form $3k + 1$, $3k + 2$ or $3k$, respectively. Since $1995 = 3(665)$, the bug will be on point 4 after 1995 jumps.

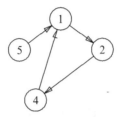

16. **(E)** Let A denote the number in attendance in Atlanta, and let B denote the number in attendance in Boston. We are given $45,000 \le A \le 55,000$ and $0.9B \le 60,000 \le 1.1B$, so $54,546 \le B \le 66,666$. Hence the largest possible difference between A and B is $66,666 - 45,000 = 21,666$, so the correct choice is **(E)**.

17. **(E)** Let O be the center of the circle. Since the sum of the interior angles in any n-gon is $(n-2)180°$, the sum of the angles in $ABCDO$ is $540°$. Since $\angle ABC = \angle BCD = 108°$ and $\angle OAB = \angle ODC = 90°$, it follows that the measure of $\angle AOD$, and thus the measure of minor arc AD, equals $144°$.

OR

Draw $\overline{AD}$. Since $\triangle AED$ is isosceles with $\angle AED = 108°$, it follows that $\angle EDA = \angle EAD = 36°$. Consequently, $\angle ADC = 108° - 36° = 72°$. Since $\angle ADC$ is a tangent-chord angle for the arc in question, the measure of the arc is $2(72°) = 144°$.

OR

Let O be the center of the circle, and extend $\overline{DC}$ and $\overline{AB}$ to meet at F. Since $\angle DCB = 108°$ and $\triangle BCF$ is isosceles, it follows that $\angle AFD = [180° - 2(180° - 108°)] = 36°$. Since $\angle ODF = \angle OAF = 90°$, in quadrilateral $OAFD$ we have angles AOD and AFD supplementary, so the measures of angle AOD and the minor arc AD are $180° - 36° = 144°$.

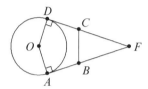

Note. A circle can be drawn tangent to two intersecting lines at given points on those lines if and only if those points are equidistant from the point of intersection of the lines.

18. **(D)** By the *Law of Sines*,

$$\frac{OB}{\sin \angle OAB} = \frac{AB}{\sin \angle AOB} = \frac{1}{1/2},$$

so $OB = 2\sin \angle OAB \leq 2\sin 90° = 2$, with equality if and only if $\angle OAB = 90°$.

OR

Consider B to be fixed on a ray originating at a variable point O, and draw another ray so the angle at O is $30°$. A possible position for A is any intersection of this ray with the circle of radius 1 centered at B. The largest value for OB for which there *is* an intersection point A occurs when $\overline{OA}$ is tangent to the circle. Since $\triangle OBA$ is a $30°$-$60°$-$90°$ triangle with $AB = 1$, it follows that $OB = 2$ is largest.

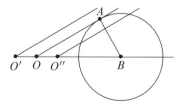

19. **(C)** Since CDE is a right triangle with $\angle C = 60°$, we have $CE = 2DC$. Also, $\angle BFD = 90° = \angle FEA$. To see that $\angle BFD = 90°$, note that

$$\angle BDF + \angle FDE + 90° = \angle BDF + 60° + 90° = 180°.$$

Thus $\angle BDF = 30°$ and since $\angle DBF = 60°$, $\angle BFD = 90°$. That $\angle FEA = 90°$ follows similarly. Since $\triangle DEF$ is equilateral, the three small triangles are congruent and $AE = DC$. Let $AC = 3x$. Then $EC = 2x$ and $DE = \sqrt{3}x$. The desired ratio is

$$\left(\frac{DE}{AC}\right)^2 = \left(\frac{\sqrt{3}x}{3x}\right)^2 = \frac{1}{3}.$$

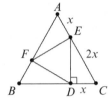

20. **(B)** The quantity $ab + c$ will be even if ab and c are both even or both odd. Furthermore, ab will be odd only when both a and b are odd, so the probability of ab being odd is $(3/5)(3/5) = 9/25$. Thus the probability of ab being even is $1 - 9/25 = 16/25$. Hence, the required probability is $(16/25) \cdot (2/5) + (9/25) \cdot (3/5) = 59/125$.

21. **(E)** The diagonals of a rectangle are of the same length and bisect each other. The given diagonal has length $\sqrt{(-4-4)^2 + (-3-3)^2} = 10$ and midpoint $(0,0)$. The other diagonal must have end points on the circle of radius 5 centered at the origin and must have integer coordinates for each end point. We must find integer solutions to $x^2 + y^2 = 5^2$. The only possible diagonals, other than the given diagonal, are the segments:

$$\overline{(0,5)(0,-5)}, \quad \overline{(5,0)(-5,0)}, \quad \overline{(3,4)(-3,-4)},$$

$$\overline{(-3,4)(3,-4)}, \quad \text{and} \quad \overline{(4,-3)(-4,3)}.$$

Each of these five, with the original diagonal, determines a rectangle.

22. **(E)** Let the sides of the pentagon be a, b, c, d and e, and let r and s be the legs of the triangular region cut off as shown. The equation $r^2 + s^2 = e^2$ has no solution in positive integers when $e = 19$ or $e = 31$. Therefore, e equals 13, 20 or 25, and the possibilities for $\{r, s, e\}$ are the well-known Pythagorean triples

$$\{5, 12, 13\}, \quad \{12, 16, 20\}, \quad \{15, 20, 25\}, \quad \{7, 24, 25\}.$$

Since 16, 15 and 24 do not appear among any of the pairwise differences of $\{13, 19, 20, 25, 31\}$, the only possibility is $\{5, 12, 13\}$. Then $a = 19$, $b = 25$, $c = 31$, $d = 20$ and $e = 13$. Hence, the area of the pentagon is $31 \times 25 - \frac{1}{2}(12 \times 5) = 775 - 30 = 745$.

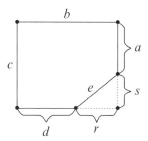

23. **(D)** Since the longest side of a triangle must be less than the sum of the other two sides, it follows that $4 < k < 26$. For the triangle to be obtuse, either $11^2 + 15^2 < k^2$, or $11^2 + k^2 < 15^2$. Therefore the 13 suitable values of k are 5, 6, 7, 8, 9, 10, 19, 20, 21, 22, 23, 24 and 25.

24. **(A)** Note that $C = A \log_{200} 5 + B \log_{200} 2 = \log_{200} 5^A + \log_{200} 2^B = \log_{200}(5^A \cdot 2^B)$, so $200^C = 5^A \cdot 2^B$. Therefore, $5^A \cdot 2^B = 200^C = (5^2 \cdot 2^3)^C = 5^{2C} 2^{3C}$. By uniqueness of prime factorization, $A = 2C$ and $B = 3C$. Letting $C = 1$ we get $A = 2$, $B = 3$ and $A + B + C = 6$. The triplet $(A, B, C) = (2, 3, 1)$ is the only solution with no common factor greater than 1. **Note.** The uniqueness is guaranteed by the *Fundamental Theorem of Arithmetic*.

25. **(B)** Since the median and mode are both 8 and the range is 18, the list must take on one of these two forms:

$$(I): \quad a, b, 8, 8, a+18 \quad \text{where } a \leq b \leq 8 \leq a + 18$$
$$\text{or} \quad (II): \quad c, 8, 8, d, c+18 \quad \text{where } c \leq 8 \leq d \leq c + 18.$$

The sum of the five integers must be 60, since their mean is 12. In case (I), the requirement that $2a + b + 34 = 60$ contradicts $a, b \leq 8$. In case (II), $2c + d + 34 = 60$ and $c \leq 8 \leq d \leq c + 18$ lead to these six pairs, (c, d):

$$(8, 10), \ (7, 12), \ (6, 14), \ (5, 16), \ (4, 18), \ (3, 20).$$

Thus, the second largest entry in the list can be any of the six numbers $d = 10, 12, 14, 16, 18, 20$.

OR

Let a be the smallest element, d the largest, and b, b and c the remaining elements. Then $b = 8$, $a + 2b + c + d = 60$, and $d - a = 18$. Solving these equations simultaneously yields $a = 13 - \frac{c}{2}$ and $d = 31 - \frac{c}{2}$. Thus for a and d to be positive integers, c must be even and ≤ 24. Further, $a \leq 8$ implies $c \geq 10$, and $c \leq d$ implies $c \leq 20$. Thus c must be one of the six integers $10, 12, 14, 16, 18, 20$. Each of these six values for c yields a list of five positive integers with the required properties.

26. **(C)** Draw segment $\overline{FC}$. Angle CFD is a right angle since arc CFD is a semicircle. Then right triangles DOE and DFC are similar, so

$$\frac{DO}{DF} = \frac{DE}{DC}.$$

Let $DO = r$ and $DC = 2r$. Substituting, we have

$$\frac{r}{8} = \frac{6}{2r}, \quad 2r^2 = 48, \quad r^2 = 24.$$

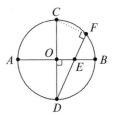

Then the area of the circle is $\pi r^2 = 24\pi$.

OR

Let $OA = OB = r$ and $OE = x$. Substituting into $AE \cdot EB = DE \cdot EF$ gives $(r+x)(r-x) = 6 \cdot 2$ so $r^2 - x^2 = 12$. In right triangle EOD, $r^2 + x^2 = 36$. Add to find $2r^2 = 48$. Thus, the area of the circle is $\pi r^2 = 24\pi$.

OR

Construct $\overline{OG} \perp \overline{DF}$ with G on $\overline{DF}$. Then $DG = DF/2 = 4$. Since $\overline{OG}$ is an altitude to the hypotenuse of right triangle EOD, we have $DE/DO = DO/DG$. Let $DO = r$. Then $6/r = r/4$, so $r^2 = 24$, and the area of the circle is $\pi r^2 = 24\pi$.

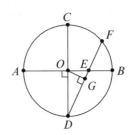

27. **(E)** Calculating the first five values of f,
$$f(1) = 0, \quad f(2) = 2, \quad f(3) = 6, \quad f(4) = 14, \quad f(5) = 30,$$
we are led to the conjecture that $f(n) = 2^n - 2$. We prove this by induction: Observe that each of the interior numbers in row n is used twice and each of the end numbers is used once as a term in computing the interior terms of row $n+1$; i.e.,
$$f(n+1) = [2f(n) - 2(n-1)] + 2n = 2f(n) + 2,$$
so if $f(n) = 2^n - 2$, then $f(n+1) = 2f(n) + 2 = 2(2^n - 2) + 2 = 2^{n+1} - 2$. Therefore, we seek the remainder when $f(100) = 2^{100} - 2$ is divided by 100. Use the fact that 76^2 has remainder 76 when divided by 100. We find
$$2^{10} = 100K + 24,$$
$$2^{20} = 100L + 76,$$
$$2^{40} = 100M + 76,$$
$$2^{80} = 100N + 76,$$
$$2^{100} = 100Q + 76,$$
for positive integers K, L, M, N, Q, so $f(100) = 2^{100} - 2$ has remainder 74 when divided by 100. **Query:** What other positive integers N have the property that N^2 has remainder N when divided by 100?

28. **(E)** Let x be the distance from the center O of the circle to the chord of length 10, and let y be the distance from O to the chord of length 14. Let r be the radius. Then,
$$x^2 + 25 = r^2,$$
$$y^2 + 49 = r^2,$$
so $x^2 + 25 = y^2 + 49.$
Therefore, $x^2 - y^2 = (x-y)(x+y) = 24.$

If the chords are on the same side of the center of the circle, $x - y = 6$. If they are on opposite sides, $x + y = 6$. But $x - y = 6$ implies that $x + y = 4$, which is impossible. Hence $x + y = 6$ and $x - y = 4$. Solve these equations simultaneously to get $x = 5$ and $y = 1$. Thus, $r^2 = 50$, and the chord parallel to the given chords and midway between them is two units from the center. If the chord is of length $2d$, then $d^2 + 4 = 50$, $d^2 = 46$, and $a = (2d)^2 = 184$.

OR

The diameter perpendicular to the chords is divided by the chord of length $\sqrt{a}$ into segments with lengths c and d as shown. Then

$$cd = \left(\frac{\sqrt{a}}{2}\right)^2 = \frac{a}{4}.$$

Treat the chords 3 units above and 3 units below similarly:

$$(c-3)(d+3) = \left(\frac{14}{2}\right)^2$$

$$(c+3)(d-3) = \left(\frac{10}{2}\right)^2.$$

Adding the last two equations, we get $2cd - 18 = 49 + 25 = 74$. Thus, $2cd = 92$ so $a = 4cd = 184$.

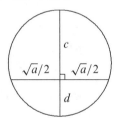

29. **(C)** Since the three factors, a, b and c, must be distinct, we seek the number of positive integer solutions to

$$abc = 2310 = 2 \cdot 3 \cdot 5 \cdot 7 \cdot 11, \quad \text{with} \quad a < b < c.$$

The prime factors of a, b and c must be disjoint subsets of $S = \{2, 3, 5, 7, 11\}$, no more than one subset can be empty, and the union of the subsets must be S. The numbers of elements in the subsets can be: $0, 1, 4$; $0, 2, 3$; $1, 1, 3$; or $1, 2, 2$. In the $0, 1, 4$ case, there are five ways to choose three subsets with these sizes. In the $0, 2, 3$ case, there are $\binom{5}{2} = 10$ ways to choose the three subsets. In the $1, 1, 3$ case, there are $\binom{5}{3} = 10$ ways to choose the three subsets. In the $1, 2, 2$ case, there are five ways of choosing the one-element subset and $\frac{1}{2} \cdot \binom{4}{2} = 3$ ways of dividing the remaining four elements into two subsets of two elements each, yielding 15 ways of choosing the three subsets in this case.

Thus there are a total of $5 + 10 + 10 + 15 = 40$ ways of choosing our three subsets and, therefore, 40 ways of expressing 2310 in the required manner. Since factorization into primes is unique, these 40 triplets of sets give distinct solutions.

OR

There are $3^5 = 243$ ordered triples, (a, b, c), of integers such that $abc = 2310$, since each of the five prime factors of $2310 = 2 \cdot 3 \cdot 5 \cdot 7 \cdot 11$ divides exactly one of a, b or c. In three of these 243 ordered triples, two of a, b, c equal 1. In the remaining 240 ordered triples, a, b and c are distinct, since 2310 is square-free. Each unordered triple whose product is 2310 is represented by $3! = 6$ of the 240 ordered triples (a, b, c), so the answer is $240/6 = 40$.

30. **(D)** Suppose the coordinates of the vertices of the unit cubes occur at (i, j, k) for all $i, j, k \in \{0, 1, 2, 3\}$. The equation of the plane that bisects the large cube's diagonal from $(0, 0, 0)$ to $(3, 3, 3)$ is $x + y + z = 9/2$. That plane meets a unit cube if and only if the ends of the unit cube's diagonal from (i, j, k) to $(i+1, j+1, k+1)$ lie on opposite sides of the plane. Therefore, this problem is equivalent to counting the number of the 27 triples (i, j, k) with $i, j, k \in \{0, 1, 2\}$ for which $i + j + k < 4.5 < i + j + k + 3$. Only eight of these 27 triples do not satisfy these inequalities:

$$(0, 0, 0), \quad (1, 0, 0), \quad (0, 1, 0), \quad (0, 0, 1),$$
$$(1, 2, 2), \quad (2, 1, 2), \quad (2, 2, 1), \quad (2, 2, 2).$$

Therefore, $27 - 8 = 19$ of the unit cubes are intersected by the plane.

A sketch can help you visualize the 19 unit cubes intersected by the plane. Suppose the plane is perpendicular to the interior diagonal $\overline{AB}$ at its midpoint. That plane intersects the surface of the large cube in a regular hexagon.

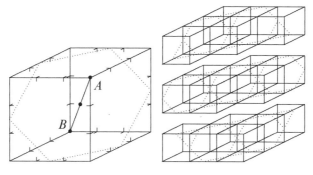

The sketch shows that nineteen of the twenty-seven unit cubes are intersected by this plane, with six each in the bottom and top layers and seven in the middle layer. The corner unit cube at vertex A and the three unit cubes adjacent to it are missed by this plane, as are the four symmetric to these at vertex B.

47th AHSME solutions, 1996

1. **(D)** The mistake occurs in the tens column where any of the digits of the addends can be decreased by 1 or the 5 in the sum changed to 6 to make the addition correct. The largest of these digits is 7.

2. **(A)** Walter gets an extra $2 per day for doing chores exceptionally well. If he never did them exceptionally well, he would get $30 for 10 days of chores. The extra $6 must be for 3 days of exceptional work.

3. **(E)** $\dfrac{(3!)!}{3!} = \dfrac{6!}{3!} = 6 \cdot 5 \cdot 4 = 120.$

   OR

   $\dfrac{(3!)!}{3!} = \dfrac{6!}{6} = 5! = 5 \cdot 4 \cdot 3 \cdot 2 = 120.$

4. **(D)** The largest possible median will occur when the three numbers not given are larger than those given. Let a, b, and c denote the three missing numbers, where $9 \leq a \leq b \leq c$. Ranked from smallest to largest, the list is

   $$3, 5, 5, 7, 8, 9, a, b, c,$$

   so the median value is 8.

5. **(E)** The largest fraction is the one with largest numerator and smallest denominator. Choice **(E)** has both.

83

6. **(E)** Since $0^z = 0$ for any $z > 0$, $f(0) = f(-2) = 0$. Since $(-1)^0 = 1$,

$$f(0) + f(-1) + f(-2) + f(-3) = (-1)^0 (1)^2 + (-3)^{-2}(-1)^0$$
$$= 1 + \frac{1}{(-3)^2}$$
$$= \frac{10}{9}.$$

7. **(B)** The sum of the children's ages is 10 because $9.45 - \$4.95 = \$4.50 = 10 \times \$0.45$. If the twins were 3 years old or younger, then the third child would not be the youngest. If the twins are 4, the youngest child must be 2.

8. **(D)** Since $3 = k \cdot 2^r$ and $15 = k \cdot 4^r$, we have

$$5 = \frac{15}{3} = \frac{k \cdot 4^r}{k \cdot 2^r} = \frac{2^{2r}}{2^r} = 2^r.$$

Thus, by definition, $r = \log_2 5$.

9. **(B)** Since line segment AD is perpendicular to the plane of PAB, angle PAD is a right angle. In right triangle PAD, $PA = 3$ and $AD = AB = 5$. By the *Pythagorean Theorem*, $PD = \sqrt{3^2 + 5^2} = \sqrt{34}$. The fact that $PB = 4$ was not needed.

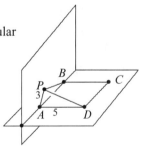

10. **(D)** There are 12 edges, 12 face diagonals, and 4 space diagonals for a total of $12 + 12 + 4 = 28$.

OR

Each pair of vertices of the cube determines a line segment. There are $\binom{8}{2} = \frac{8!}{(8-2)!2!} = \frac{8 \cdot 7}{2} = 28$ such pairs.

11. **(D)** The endpoints of each of these line segments are at distance $\sqrt{2^2 + 1^2} = \sqrt{5}$ from the center of the circle. The region is therefore an annulus with inner radius 2 and outer radius $\sqrt{5}$. The area covered is $\pi(\sqrt{5})^2 - \pi(2)^2 = \pi$.

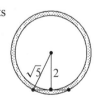

Note. The area of the annular region covered by the segments of length 2 does not depend on the radius of the circle.

12. **(B)** Since k is odd, $f(k) = k + 3$. Since $k + 3$ is even,
$$f(f(k)) = f(k+3) = (k+3)/2.$$
If $(k+3)/2$ is odd, then
$$27 = f(f(f(k))) = f((k+3)/2) = (k+3)/2 + 3,$$
which implies that $k = 45$. This is not possible because $f(f(f(45))) = f(f(48)) = f(24) = 12$. Hence $(k+3)/2$ must be even, and
$$27 = f(f(f(k))) = f((k+3)/2) = (k+3)/4,$$
which implies that $k = 105$. Checking, we find that
$$f(f(f(105))) = f(f(108)) = f(54) = 27.$$
Hence the sum of the digits of k is $1 + 0 + 5 = 6$.

13. **(D)** Let x be the number of meters that Moonbeam runs to overtake Sunny, and let r and mr be the rates of Sunny and Moonbeam, respectively. Because Sunny runs $x - h$ meters in the same time that Moonbeam runs x meters, it follows that $x - h/r = x/mr$. Solving for x, we get $x = hm/m - 1$.

14. **(C)** Since $E(100) = E(00)$, the result is the same as $E(00) + E(01) + E(02) + E(03) + \cdots + E(99)$, which is the same as
$$E(00010203\ldots99).$$
There are 200 digits, and each digit occurs 20 times, so the sum of the even digits is $20(0 + 2 + 4 + 6 + 8) = 20(20) = 400$.

15. **(B)** Let the base of the rectangle be b and the height a. Triangle A has an altitude of length $b/2$ to a base of length a/n, and triangle B has an altitude of length $a/2$ to a base of length b/m. Thus the required ratio of areas is
$$\frac{\frac{1}{2} \cdot \frac{a}{n} \cdot \frac{b}{2}}{\frac{1}{2} \cdot \frac{b}{m} \cdot \frac{a}{2}} = \frac{m}{n}.$$

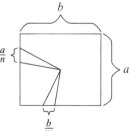

16. **(D)** There are 15 ways in which the third outcome is the sum of the first two outcomes.

(1,1,2)	(2,1,3)	(3,1,4)	(4,1,5)	(5,1,6)
(1,2,3)	(2,2,4)	(3,2,5)	(4,2,6)	
(1,3,4)	(2,3,5)	(3,3,6)		
(1,4,5)	(2,4,6)			
(1,5,6)				

Since the three tosses are independent, all of the 15 possible outcomes are equally likely. At least one "2" appears in exactly eight of these outcomes, so the required probability is 8/15.

17. **(E)** In the 30°-60°-90° triangle CEB, $BC = 6\sqrt{3}$. Therefore, $FD = AD - AF = 6\sqrt{3} - 2$. In the 30°-60°-90° triangle CFD, $CD = FD\sqrt{3} = 18 - 2\sqrt{3}$. The area of rectangle $ABCD$ is

$$(BC)(CD) = \left(6\sqrt{3}\right)\left(18 - 2\sqrt{3}\right) = 108\sqrt{3} - 36 \approx 151.$$

18. **(D)** Let D and F denote the centers of the circles. Let C and B be the points where the x-axis and y-axis intersect the tangent line, respectively. Let E and G denote the points of tangency as shown. We know that $AD = DE = 2$, $DF = 3$, and $FG = 1$. Let $FC = u$ and $AB = y$. Triangles FGC and DEC are similar, so

$$\frac{u}{1} = \frac{u+3}{2},$$

which yields $u = 3$. Hence, $GC = \sqrt{8}$. Also, triangles BAC and FGC are similar, which yields

$$\frac{y}{1} = \frac{BA}{FG} = \frac{AC}{GC} = \frac{8}{\sqrt{8}} = \sqrt{8} = 2\sqrt{2}.$$

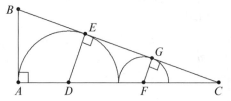

19. **(D)** Let R and S be the vertices of the smaller hexagon adjacent to vertex E of the larger hexagon, and let O be the center of the hexagons. Then, since $\angle ROS = 60°$, quadrilateral $ORES$ encloses 1/6 of the

area of $ABCDEF$, $\triangle ORS$ encloses 1/6 of the area of the smaller hexagon, and $\triangle ORS$ is equilateral. Let T be the center of $\triangle ORS$. Then triangles TOR, TRS, and TSO are congruent isosceles triangles with largest angle 120°. Triangle ERS is an isosceles triangle with largest angle 120° and a side in common with $\triangle TRS$, so $ORES$ is partitioned into four congruent triangles, exactly three of which form $\triangle ORS$. Since the ratio of the area enclosed by the small regular hexagon to the area of $ABCDEF$ is the same as the ratio of the area enclosed by $\triangle ORS$ to the area enclosed by $ORES$, the ratio is 3/4.

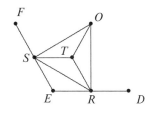

OR

Let M and N denote the midpoints of $\overline{AB}$ and $\overline{AF}$, respectively. Then $MN = AM\sqrt{3}$ since $\triangle AMO$ is a 30°-60°-90° triangle and $MN = MO$. It follows that the hexagons are similar, with similarity ratio $\frac{1}{2}\sqrt{3}$. Thus the desired quotient is $(\frac{1}{2}\sqrt{3})^2 = \frac{3}{4}$.

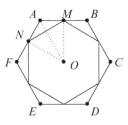

20. **(C)** Let $O = (0,0)$, $P = (6,8)$, and $Q = (12,16)$. As shown in the figure, the shortest route consists of tangent $\overline{OT}$, minor arc TR, and tangent $\overline{RQ}$. Since $OP = 10$, $PT = 5$, and $\angle OTP$ is a right angle, it follows that $\angle OPT = 60°$ and $OT = 5\sqrt{3}$. By similar reasoning, $\angle QPR = 60°$ and $QR = 5\sqrt{3}$. Because O, P, and Q are collinear (why?), $\angle RPT = 60°$, so arc TR is of length $5\pi/3$. Hence the length of the shortest route is $2\left(5\sqrt{3}\right) + 5\pi/3$.

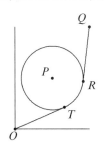

21. **(D)** Let $\angle ABD = x$ and $\angle BAC = y$. Since triangles ABC and ABD are isosceles, $\angle C = (180° - y)/2$ and $\angle D = (180° - x)/2$. Then, noting that $x + y = 90°$, we have $\angle C + \angle D = (360° - (x + y))/2 = 135°$.

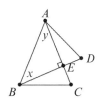

OR

Consider the interior angles of pentagon $ADECB$. Since triangles ABC and ABD are isosceles, $\angle C = \angle B$ and $\angle D = \angle A$. Since $\overline{BD} \perp \overline{AC}$, the interior angle at E measures $270°$. Since $540°$ is the sum of the interior angles of any pentagon,

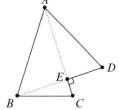

$$\angle A + \angle B + \angle C + \angle D + \angle E$$
$$= 2\angle C + 2\angle D + 270° = 540°,$$

from which it follows that

$$\angle C + \angle D = 135°.$$

22. **(B)** Because all quadruples are equally likely, we need only examine the six clockwise orderings of the points: $ACBD$, $ADBC$, $ABCD$, $ADCB$, $ABDC$, and $ACDB$. Only the first two of these equally likely orderings satisfy the intersection condition, so the probability is $2/6 = 1/3$.

23. **(B)** Let a, b, and c be the dimensions of the box. It is given that

$$140 = 4a + 4b + 4c \quad \text{and} \quad 21 = \sqrt{a^2 + b^2 + c^2},$$

hence

$$35 = a + b + c \quad (1) \quad \text{and} \quad 441 = a^2 + b^2 + c^2 \quad (2).$$

Square both sides of (1) and combine with (2) to obtain

$$1225 = (a + b + c)^2$$
$$= a^2 + b^2 + c^2 + 2ab + 2bc + 2ca$$
$$= 441 + 2ab + 2bc + 2ca.$$

Thus the surface area of the box is $2ab + 2bc + 2ca = 1225 - 441 = 784$.

24. **(B)** The kth 1 is at position

$$1 + 2 + 3 + \cdots + k = \frac{k(k+1)}{2}$$

and $\frac{49(50)}{2} < 1234 < \frac{50(51)}{2}$, so there are 49 1's among the first 1234 terms. All the other terms are 2's, so the sum is $1234(2) - 49 = 2419$.

OR

The sum of all the terms through the occurrence of the kth 1 is

$$1 + (2+1) + (2+2+1) + \cdots + (\underbrace{2+2+\cdots+2}_{k-1}+1)$$

$$= 1 + 3 + 5 + \cdots + (2k-1)$$

$$= k^2.$$

The kth 1 is at position

$$1 + 2 + 3 + \cdots + k = \frac{k(k+1)}{2}.$$

It follows that the last 1 among the first 1234 terms of the sequence occurs at position 1225 for $k = 49$. Thus, the sum of the first 1225 terms is $49^2 = 2401$, and the sum of the next nine terms, all of which are 2's, is 18, for a total of $2401 + 18 = 2419$.

25. **(B)** The equation $x^2 + y^2 = 14x + 6y + 6$ can be written

$$(x-7)^2 + (y-3)^2 = 8^2,$$

which defines a circle of radius 8 centered at $(7, 3)$. If k is a possible value of $3x + 4y$ for (x, y) on the circle, then the line $3x + 4y = k$ must intersect the circle in at least one point. The largest value of k occurs when the line is tangent to the circle, and is therefore perpendicular to the radius at the point of tangency. Because the slope of the tangent line is $-3/4$, the slope of the radius is $4/3$. It follows that the point on the circle that yields the maximum value of $3x + 4y$ is one of the two points of tangency,

$$x = 7 + \frac{3 \cdot 8}{5} = \frac{59}{5}, \qquad y = 3 + \frac{4 \cdot 8}{5} = \frac{47}{5},$$

or

$$x = 7 - \frac{3 \cdot 8}{5} = \frac{11}{5}, \qquad y = 3 - \frac{4 \cdot 8}{5} = -\frac{17}{5}.$$

The first point of tangency gives

$$3x + 4y = 3 \cdot \frac{59}{5} + 4 \cdot \frac{47}{5} = \frac{177}{5} + \frac{188}{5} = 73,$$

and the second one gives

$$3x + 4y = \frac{33}{5} - \frac{68}{5} = -7.$$

Thus 73 is the desired maximum, while -7 is the minimum.

OR

Suppose that $k = 3x + 4y$ is a possible value. Substituting $y = (k-3x)/4$ into $x^2 + y^2 = 14x + 6y + 6$, we get $16x^2 + (k-3x)^2 = 224x + 24(k-3x) + 96$, which simplifies to

$$25x^2 - 2(3k + 76)x + (k^2 - 24k - 96) = 0. \qquad (1)$$

If the line $3x + 4y = k$ intersects the given circle, the discriminant of (1) must be nonnegative. Thus we get $(3k + 76)^2 - 25(k^2 - 24k - 96) \geq 0$, which simplifies to

$$(k - 73)(k + 7) \leq 0.$$

Hence $-7 \leq k \leq 73$.

26. **(B)** The hypothesis of equally likely events can be expressed as

$$\frac{\binom{r}{4}}{\binom{n}{4}} = \frac{\binom{r}{3}\binom{w}{1}}{\binom{n}{4}} = \frac{\binom{r}{2}\binom{w}{1}\binom{b}{1}}{\binom{n}{4}} = \frac{\binom{r}{1}\binom{w}{1}\binom{b}{1}\binom{g}{1}}{\binom{n}{4}}$$

where r, w, b, and g denote the number of red, white, blue, and green marbles, respectively, and $n = r + w + b + g$. Eliminating common terms and solving for r in terms of w, b, and g, we get

$$r - 3 = 4w, \qquad r - 2 = 3b, \qquad \text{and} \qquad r - 1 = 2g.$$

The smallest r for which w, b, and g are all positive integers is $r = 11$, with corresponding values $w = 2$, $b = 3$, and $g = 5$. So the smallest total number of marbles is $11 + 2 + 3 + 5 = 21$.

27. **(D)** From the description of the first ball we find that $z \geq 9/2$, and from that of the second, $z \leq 11/2$. Because z must be an integer, the only possible lattice points in the intersection are of the form $(x, y, 5)$. Substitute $z = 5$ into the inequalities defining the balls:

$$x^2 + y^2 + \left(z - \frac{21}{2}\right)^2 \leq 6^2 \qquad \text{and} \qquad x^2 + y^2 + (z-1)^2 \leq \left(\frac{9}{2}\right)^2.$$

These yield

$$x^2 + y^2 + \left(-\frac{11}{2}\right)^2 \leq 6^2 \qquad \text{and} \qquad x^2 + y^2 + (4)^2 \leq \left(\frac{9}{2}\right)^2,$$

which reduce to

$$x^2 + y^2 \le \frac{23}{4} \quad \text{and} \quad x^2 + y^2 \le \frac{17}{4}.$$

If $(x, y, 5)$ satisfies the second inequality, then it must satisfy the first one. The only remaining task is to count the lattice points that satisfy the second inequality. There are 13:

$(-2, 0, 5)$, $(2, 0, 5)$, $(0, -2, 5)$, $(0, 2, 5)$, $(-1, -1, 5)$,
$(1, -1, 5)$, $(-1, 1, 5)$, $(1, 1, 5)$, $(-1, 0, 5)$, $(1, 0, 5)$,
$(0, -1, 5)$, $(0, 1, 5)$, and $(0, 0, 5)$.

28. **(C)** Let h be the required distance. Find the volume of pyramid $ABCD$ as a third of the area of a triangular base times the altitude to that base in two different ways, and equate these volumes. Use the altitude $\overline{AD}$ to $\triangle BCD$ to find that the volume is 8. Next, note that h is the length of the altitude of the pyramid from D to $\triangle ABC$. Since the sides of $\triangle ABC$ are 5, 5, and $4\sqrt{2}$, by the Pythagorean Theorem the altitude to the side of length $4\sqrt{2}$ is $a = \sqrt{17}$. Thus, the area of $\triangle ABC$ is $2\sqrt{34}$, and the volume of the pyramid is $2\sqrt{34}h/3$. Equating the volumes yields

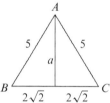

$$2\sqrt{34}h/3 = 8, \quad \text{and thus} \quad h = 12/\sqrt{34} \approx 2.1.$$

OR

Imagine the parallelepiped embedded in a coordinate system as shown in the diagram. The equation for the plane (in intercept form) is $\frac{x}{4} + \frac{y}{4} + \frac{z}{3} = 1$. Thus, it can be expressed as $3x + 3y + 4z - 12 = 0$. The formula for the distance d from a point (a, b, c) to the plane $Rx + Sy + Tz + U = 0$ is given by

$$d = \frac{|Ra + Sb + Tc + U|}{\sqrt{R^2 + S^2 + T^2}},$$

which in this case is

$$\frac{|-12|}{\sqrt{3^2 + 3^2 + 4^2}} = \frac{12}{\sqrt{34}} \approx 2.1.$$

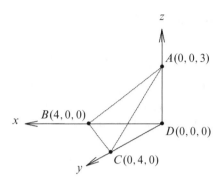

29. **(C)** Let $2^{e_1} 3^{e_2} 5^{e_3} \cdots$ be the prime factorization of n. Then the number of positive divisors of n is $(e_1 + 1)(e_2 + 1)(e_3 + 1) \cdots$. In view of the given information, we have

$$28 = (e_1 + 2)(e_2 + 1)P \quad \text{and} \quad 30 = (e_1 + 1)(e_2 + 2)P,$$

where $P = (e_3 + 1)(e_4 + 1) \cdots$. Subtracting the first equation from the second, we obtain $2 = (e_1 - e_2)P$, so either $e_1 - e_2 = 1$ and $P = 2$, or $e_1 - e_2 = 2$ and $P = 1$. The first case yields $14 = (e_1 + 2)e_1$ and $(e_1 + 1)^2 = 15$; since e_1 is a nonnegative integer, this is impossible. In the second case, $e_2 = e_1 - 2$ and $30 = (e_1 + 1)e_1$, from which we find $e_1 = 5$ and $e_2 = 3$. Thus $n = 2^5 3^3$, so $6n = 2^6 3^4$ has $(6 + 1)(4 + 1) = 35$ positive divisors.

30. **(E)** In hexagon $ABCDEF$, let $AB = BC = CD = 3$ and let $DE = EF = FA = 5$. Since arc BAF is one third of the circumference of the circle, it follows that $\angle BCF = \angle BEF = 60°$. Similarly, $\angle CBE = \angle CFE = 60°$. Let P be the intersection of $\overline{BE}$ and $\overline{CF}$, Q that of $\overline{BE}$ and $\overline{AD}$, and R that of $\overline{CF}$ and $\overline{AD}$. Triangles EFP and BCP are equilateral, and by symmetry, triangle PQR is isosceles and thus also equilateral. Furthermore, $\angle BAD$ and $\angle BED$ subtend the same arc, as do $\angle ABE$ and $\angle ADE$. Hence triangles ABQ and EDQ are similar. Therefore,

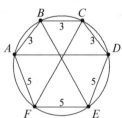

$$\frac{AQ}{EQ} = \frac{BQ}{DQ} = \frac{AB}{ED} = \frac{3}{5}.$$

It follows that

$$\frac{\frac{AD-PQ}{2}}{PQ+5} = \frac{3}{5} \quad \text{and} \quad \frac{3-PQ}{\frac{AD+PQ}{2}} = \frac{3}{5}.$$

Solving the two equations simultaneously yields $AD = 360/49$, so $m + n = 409$.

OR

In hexagon $ABCDEF$, let $AB = BC = CD = a$ and let $DE = EF = FA = b$. Let O denote the center of the circle, and let r denote the radius. Since the arc BAF is one-third of the circle, it follows that $\angle BAF = \angle FOB = 120°$. By using the Law of Cosines to compute BF two ways, we have $a^2 + ab + b^2 = 3r^2$. Let $\angle AOB = 2\theta$. Then $a = 2r\sin\theta$, and

$$AD = 2r\sin(3\theta)$$

$$= 2r\sin\theta \cdot (3 - 4\sin^2\theta)$$

$$= a\left(3 - \frac{a^2}{r^2}\right)$$

$$= 3a\left(1 - \frac{a^2}{a^2 + ab + b^2}\right)$$

$$= \frac{3ab(a + b)}{a^2 + ab + b^2}.$$

Substituting $a = 3$ and $b = 5$, we get $AD = 360/49$, so $m + n = 409$.

OR

In hexagon $ABCDEF$, let $AB = BC = CD = 3$ and minor arcs AB, BC, and CD each be $x°$, let $DE = EF = FA = 5$ and minor arcs DE, EF, and FA each be $y°$. Then

$$3x° + 3y° = 360°, \quad \text{so} \quad x° + y° = 120°.$$

Therefore, $\angle BAF = 120°$, so $BF^2 = 3^2 + 5^2 - 2\cdot 3\cdot 5\cos 120° = 49$ by the Law of Cosines, so $BF = 7$. Similarly, $CE = 7$. Using Ptolemy's Theorem in quadrilateral $BCEF$, we have

$$BE \cdot CF = CF^2 = 15 + 49 = 64 \quad \text{so} \quad CF = 8.$$

Using Ptolemy's Theorem in quadrilateral $ABCF$, we find $AC = 39/7$. Finally, using Ptolemy's Theorem in quadrilateral $ABCD$, we have $AC^2 = 3(AD) + 9$ and, since $AC = 39/7$, we have $AD = 360/49$ and $m + n = 409$.

Note. *Ptolemy's Theorem:* If a quadrilateral is inscribed in a circle, the product of the diagonals equals the sum of the products of the opposite sides.

48th AHSME solutions, 1997

1. **(C)** Since a × 3 has units digit 9, a must be 3. Hence b × 3 has units digit 2, so b must be 4. Thus, a + b = 7.

2. **(D)** Each polygon in the sequence below has the same perimeter, which is 44.

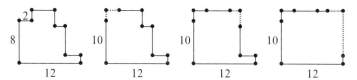

3. **(D)** Since each summand is nonnegative, the sum is zero only when each term is zero. Hence the only solution is $x = 3$, $y = 4$, and $z = 5$, so the desired sum is 12.

4. **(A)** If a is 50% larger than c, then $a = 1.5c$. If b is 25% larger than c, then $b = 1.25c$. So $a/b = 1.5c/1.25c = 6/5 = 1.20$, and $a = 1.20b$. Therefore, a is 20% larger than b.

5. **(C)** Let x and y denote the width and height of one of the five rectangles, with $x < y$. Then $5x + 4y = 176$ and $3x = 2y$. Solve simultaneously to get $x = 16$ and $y = 24$. The perimeter in question is $2 \cdot 16 + 2 \cdot 24 = 80$.

6. **(B)** The 200 terms can be grouped into 100 odd-even pairs, each with a sum of -1. Thus the sum of the first 200 terms is $-1 \cdot 100 = -100$, and the average of the first 200 terms is $-100/200 = -0.5$.

7. **(D)** Not all seven integers can be larger than 13. If six of them were each 14, then the seventh could be $-(6 \times 14) - 1$, so that the sum would be -1.

8. **(D)** The cost of 25 books is $C(25) = 25 \times \$11 = \$275$. The cost of 24 books is $C(24) = 24 \times \$12 = \$288$, while 23 and 22 books cost $C(23) = 23 \times \$12 = \$276$ and $C(22) = 22 \times \$12 = \$264$, respectively. Thus it is cheaper to buy 25 books than 23 or 24 books. Similarly, 49 books cost less than 45, 46, 47, or 48 books. In these six cases the total cost is reduced by ordering more books. There are no other cases.

OR

A discount of $1 per book is given on orders of at least 25 books. This discount is larger than $2 \times \$12$, the cost of two books at the regular price. Thus, $n = 23$ and $n = 24$ are two values of n for which it is cheaper to order more books. Similarly, we receive an additional $1 discount per book when we buy at least 49 books. This discount would enable us to buy four more books at $10 per book, so there are four more values of n: 45, 46, 47, and 48, for a total of six values.

9. **(C)** In right triangle BAE, $BE = \sqrt{2^2 + 1^2} = \sqrt{5}$. Since $\triangle CFB \sim \triangle BAE$, it follows that

$$[CFB] = (CB/BE)^2 \cdot [BAE] = \left(\frac{2}{\sqrt{5}}\right)^2 \cdot \left(\frac{1}{2}\right)(2 \cdot 1) = \frac{4}{5}.$$

Then

$$[CDEF] = [ABCD] - [BAE] - [CFB] = 4 - 1 - \frac{4}{5} = \frac{11}{5}.$$

OR

Draw the figure in the plane as shown with B at the origin. An equation of the line BE is $y = 2x$, and, since the lines are perpendicular, an equation of the line CF is $y = -(x-2)/2$. Solve these two equations simultaneously to get $F = (2/5, 4/5)$ and

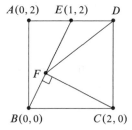

$$[CDEF] = [DEF] + [CDF] = \frac{1}{2}(1)\left(2 - \frac{4}{5}\right) + \frac{1}{2}(2)\left(2 - \frac{2}{5}\right) = \frac{11}{5}.$$

10. **(D)** There are 36 equally likely outcomes as shown in the following table.

(1, 1)	(1,2)	(1,4)	(1,4)	(1, 5)	(1,6)
(2,1)	(2, 2)	(2, 4)	(2, 4)	(2,5)	(2, 6)
(3, 1)	(3,2)	(3,4)	(3,4)	(3, 5)	(3,6)
(3, 1)	(3,2)	(3,4)	(3,4)	(3, 5)	(3,6)
(5, 1)	(5,2)	(5,4)	(5,4)	(5, 5)	(5,6)
(6,1)	(6, 2)	(6, 4)	(6, 4)	(6,5)	(6, 6)

Exactly 20 of the outcomes have an odd sum. Therefore, the probability is $20/36 = 5/9$.

OR

The sum is odd if and only if one number is even and the other is odd. The probability that the first number is even and the second is odd is $(1/3) \cdot (1/3)$, and the probability that the first is odd and the second is even is $(2/3) \cdot (2/3)$. Therefore, the required probability is $(1/3)^2 + (2/3)^2 = 5/9$.

11. **(D)** The average for games six through nine was $(23 + 14 + 11 + 20)/4 = 17$, which exceeded her average for the first five games. Therefore, she scored at most $5 \cdot 17 - 1 = 84$ points in the first five games. Because her average after ten games was more than 18, she scored at least 181 points in the ten games, implying that she scored at least $181 - 84 - 68 = 29$ points in the tenth game.

12. **(E)** Since $mb > 0$, the slope and the y-intercept of the line are either both positive or both negative. In either case, the line slopes away from the positive x-axis and does not intersect it. The answer is therefore $(1997, 0)$. Note that the other four points lie on lines for which $mb > 0$. For example, $(0, 1997)$ lies on $y = x + 1997$; $(0, -1997)$ lies on $y = -x - 1997$; $(19, 97)$ lies on $y = 5x + 2$; and $(19, -97)$ lies on $y = -5x - 2$.

13. **(E)** Let $N = 10x + y$. Then $10x + y + 10y + x = 11(x + y)$ must be a perfect square. Since $1 \leq x + y \leq 18$, it follows that $x + y = 11$. There are eight such numbers: 29, 38, 47, 56, 65, 74, 83, and 92.

14. **(B)** Let x be the number of geese in 1996, and let k be the constant of proportionality. Then $x - 39 = 60k$ and $123 - 60 = kx$. Solve the second equation for k, and use that value to solve for x in the first equation, obtaining $x - 39 = 60 \cdot 63/x$. Thus $x^2 - 39x - 3780 = 0$. Factoring yields $(x - 84)(x + 45) = 0$. Since x is positive, it follows that $x = 84$.

15. **(D)** Let the medians meet at G. Then $CG = (2/3)CE = 8$ and the area of triangle BCD is $(1/2)BD \cdot CG = (1/2) \cdot 8 \cdot 8 = 32$. Since BD is a median, triangles ABD and DBC have the same area. Hence the area of the triangle is 64.

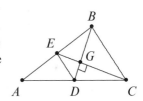

OR

Since the medians are perpendicular, the area of the quadrilateral $BCDE$ is half the product of the diagonals $(12)(8)/2 = 48$. (Why?) However, D and E are midpoints, which makes the area of triangle AED one fourth of the area of triangle ABC. Thus the area of $BCDE$ is three fourths of the area of triangle ABC. It follows that the area of triangle ABC is 64.

16. **(D)** If only three entries are altered, then either two lines are not changed at all, or some entry is the only entry in its row and the only entry in its column that is changed. In either case, at least two of the six sums remain the same. However, four alterations are enough. For example, replacing 4 by 5, 1 by 3, 2 by 7, and 6 by 9 results in the array

$$\begin{bmatrix} 5 & 9 & 7 \\ 8 & 3 & 9 \\ 3 & 5 & 7 \end{bmatrix}$$

for which the six sums are all different.

17. **(A)** The line $x = k$ intersects $y = \log_5(x + 4)$ and $y = \log_5 x$ at $(k, \log_5(k + 4))$ and $(k, \log_5 k)$, respectively. Since the length of the vertical segment is 0.5,

$$0.5 = \log_5(k + 4) - \log_5 k = \log_5 \frac{k + 4}{k},$$

so $1 + 4/k = \sqrt{5}$. Solving for k yields
$$k = \frac{4}{\sqrt{5}-1} = 1 + \sqrt{5},$$
so $a + b = 6$.

18. **(E)** When 10 is added to a number in the list, the mean increases by 2, so there must be five numbers in the original list whose sum is $5 \cdot 22 = 110$. Since 10 is the smallest number in the list and m is the median, we may assume
$$10 \leq a \leq m \leq b \leq c,$$
denoting the other members of the list by a, b, and c. Since the mode is 32, we must have $b = c = 32$; otherwise, $10 + m + a + b + c$ would be larger than 110. So $a + m = 36$. Since decreasing m by 8 decreases the median by 4, a must be 4 less than m. Solving $a + m = 36$ and $m - a = 4$ for m gives $m = 20$.

19. **(D)** Let D and E denote the points of tangency on the y- and x-axes, respectively, and let $\overline{BC}$ be tangent to the circle at F. Tangents to a circle from a point are equal, so $BE = BF$ and $CD = CF$. Let $x = BF$ and $y = CF$. Because $x + y = BC = 2$, the radius of the circle is
$$\frac{(1+x) + (\sqrt{3}+y)}{2} = \frac{3 + \sqrt{3}}{2} \approx 2.37.$$

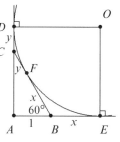

OR

Let r be the radius of the circle. The area r^2 of square $AEOD$ may also be expressed as the sum of the areas of quadrilaterals $OFBE$ and $ODCF$ and triangle ABC. This is given by $rx + ry + \sqrt{3}/2$, where $x + y = 2$. Thus
$$r^2 = 2r + \frac{\sqrt{3}}{2}.$$
Solving for r using the quadratic formula yields the positive solution
$$r = 1 + \sqrt{1 + \frac{\sqrt{3}}{2}} \approx 2.37.$$

20. **(A)** Since

$$1 + 2 + 3 + \cdots + 100 = (100)(101)/2 = 5050,$$

it follows that the sum of any sequence of 100 consecutive positive integers starting with $a + 1$ is of the form

$$(a+1) + (a+2) + (a+3) + \cdots + (a+100) =$$
$$100a + (1 + 2 + 3 + \cdots + 100) = 100a + 5050.$$

Consequently, such a sum has 50 as its rightmost two digits. Choice A is the sum of the 100 integers beginning with 16,273,800.

21. **(C)** Since $\log_8 n = \frac{1}{3}(\log_2 n)$, it follows that $\log_8 n$ is rational if and only if $\log_2 n$ is rational. The nonzero numbers in the sum will therefore be all numbers of the form $\log_8 n$, where n is an integral power of 2. The highest power of 2 that does not exceed 1997 is 2^{10}, so the sum is:

$$\log_8 1 + \log_8 2 + \log_8 2^2 + \log_8 2^3 + \cdots + \log_8 2^{10} =$$

$$0 + \frac{1}{3} + \frac{2}{3} + \frac{3}{3} + \cdots + \frac{10}{3} = \frac{55}{3}.$$

Challenge. Prove that $\log_2 3$ is irrational. Prove that, for every integer n, $\log_2 n$ is rational if and only if n is an integral power of 2.

22. **(E)** Let A, B, C, D, and E denote the amounts Ashley, Betty, Carlos, Dick, and Elgin had for shopping, respectively. Then $A - B = \pm 19$, $B - C = \pm 7$, $C - D = \pm 5$, $D - E = \pm 4$, and $E - A = \pm 11$. The sum of the left sides is zero, so the sum of the right sides must also be zero. In other words, we must choose some subset S of $\{4, 5, 7, 11, 19\}$ which has the same element-sum as its complement. Since $4 + 5 + 7 + 11 + 19 = 46$, the sum of the members of S is 23. Hence S is either the set $\{4, 19\}$ or its complement $\{5, 7, 11\}$. Thus either $A - B$ and $D - E$ are the only positive differences or $B - C$, $C - D$, and $E - A$ are. In the former case, expressing A, B, C, and D in terms of E, we get $5E + 6 = 56$, which yields $E = 10$. In the latter case, the same strategy yields $5E - 6 = 56$, which leads to non-integer values. Hence $E = 10$.

23. **(D)** The polyhedron is a unit cube with a corner cut off. The missing corner may be viewed as a pyramid whose altitude is 1 and whose

base is an isosceles right triangle (shaded in the figure). The area of the base is $1/2$. The pyramid's volume is therefore

$$\left(\frac{1}{3}\right)\left(\frac{1}{2}\right)(1) = \frac{1}{6},$$

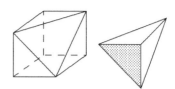

so the polyhedron's volume is

$$1 - \frac{1}{6} = \frac{5}{6}.$$

24. **(B)** The number of five-digit rising numbers that begin with 1 is $\binom{8}{4} = 70$, since the rightmost four digits must be chosen from the eight-member set $\{2, 3, 4, 5, 6, 7, 8, 9\}$, and, once they are chosen, they can be arranged in increasing order in just one way. Similarly, the next $\binom{7}{4} = 35$ integers in the list begin with 2. So the 97th integer in the list is the 27th among those that begin with 2. Among those that begin with 2, there are $\binom{6}{3} = 20$ that begin with 23 and $\binom{5}{3} = 10$ that begin with 24. Therefore, the 97th is the 7th of those that begin with 24. The first six of those beginning with 24 are 24567, 24568, 24569, 24578, 24579, 24589, and the seventh is 24678. The digit 5 is not used in the representation.

OR

As above, note that there are 105 integers in the list starting with either 1 or 2, so the 97th one is ninth from the end. Count backwards: 26789, 25789, 25689, 25679, 25678, 24789, 24689, 24679, and the ninth is 24678. Thus 5 is a missing digit.

25. **(B)** Let O be the intersection of $\overline{AC}$ and $\overline{BD}$. Then O is the midpoint of $\overline{AC}$ and $\overline{BD}$, so $\overline{OM}$ and $\overline{ON}$ are the midlines in trapezoids $ACC'A'$ and $BDD'B'$, respectively. Hence

$$OM = \frac{10 + 18}{2} = 14 \quad \text{and} \quad ON = \frac{8 + 22}{2} = 15.$$

Since $OM \parallel AA'$, $ON \parallel BB'$, and $AA' \parallel BB'$, it follows that O, M, and N are collinear. Therefore,

$$MN = |OM - ON| = |14 - 15| = 1.$$

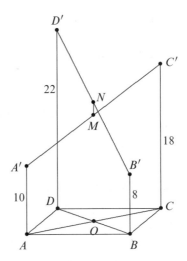

Note. In general, if $AA' = a$, $BB' = b$, $CC' = c$, and $DD' = d$, then $MN = |a - b + c - d|/2$.

26. **(A)** Construct a circle with center P and radius PA. Then C lies on the circle, since the angle ACB is half angle APB. Extend $\overline{BP}$ through P to get a diameter $\overline{BE}$. Since A, B, C, and E are concyclic,

$$AD \cdot CD = ED \cdot BD$$
$$= (PE + PD)(PB - PD)$$
$$= (3 + 2)(3 - 2)$$
$$= 5.$$

OR

Let E denote the point where $\overline{AC}$ intersects the angle bisector of angle APB. Note that $\triangle PED \sim \triangle CBD$. Hence $DE/2 = 1/DC$ so $DE \cdot DC = 2$. Apply the *Angle Bisector Theorem* to $\triangle APD$ to obtain

$$\frac{EA}{DE} = \frac{PA}{PD} = \frac{3}{2}.$$

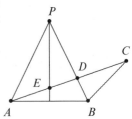

Thus $DA \cdot DC = (DE + EA) \cdot DC = (DE + 1.5 DE) \cdot DC = 2.5 \cdot DE \cdot DC = 5.$

27. **(D)** We may replace x with $x + 4$ in
$$f(x + 4) + f(x - 4) = f(x) \tag{2}$$
to get
$$f(x + 8) + f(x) = f(x + 4). \tag{3}$$
From (1) and (2), we deduce that $f(x + 8) = -f(x - 4)$. Replacing x with $x + 4$, the latter equation yields $f(x + 12) = -f(x)$. Now replacing x in this last equation with $x + 12$ yields $f(x + 24) = -f(x + 12)$. Consequently, $f(x + 24) = f(x)$ for all x, so that a least period p exists and is at most 24. On the other hand, the function $f(x) = \sin(\pi x/12)$ has fundamental period 24, and satisfies (1), so $p \geq 24$. Hence $p = 24$.

OR

Let x_0 be arbitrary, and let $y_k = f(x_0 + 4k)$ for $k = 0, 1, 2, \ldots$. Then $f(x + 4) = f(x) - f(x - 4)$ for all x implies $y_{k+1} = y_k - y_{k-1}$, so if $y_0 = a$ and $y_1 = b$, then $y_2 = b - a$, $y_3 = -a$, $y_4 = -b$, $y_5 = a - b$, $y_6 = a$, and $y_7 = b$. It follows that the sequence (y_k) is periodic with period 6 and, since x_0 was arbitrary, f is periodic with period 24. Since $f(x) = \sin(\pi x/12)$ has fundamental period 24 and satisfies $f(x + 4) + f(x - 4) = f(x)$, it follows that $p \geq 24$. Hence $p = 24$.

28. **(E)** If $c \geq 0$, then $ab - |a + b| = 78$, so $(a - 1)(b - 1) = 79$ or $(a + 1)(b + 1) = 79$. Since 79 is prime, $\{a, b\}$ is $\{2, 80\}$, $\{-78, 0\}$, $\{0, 78\}$, or $\{-80, -2\}$. Hence, $|a + b| = 78$ or $|a + b| = 82$, and from the first equation in the hypothesis, it follows that $c < 0$, a contradiction. On the other hand, if $c < 0$, then $ab + |a + b| = 116$, so $(a+1)(b+1) = 117$ or $(a-1)(b-1) = 117$. Since $117 = 3^2 \cdot 13$, we distinguish the following cases:

$$\{a, b\} = \{0, 116\} \text{ yields } c = -97;$$
$$\{a, b\} = \{2, 38\} \text{ yields } c = -21;$$
$$\{a, b\} = \{8, 12\} \text{ yields } c = -1;$$
$$\{a, b\} = \{-116, 0\} \text{ yields } c = -97;$$
$$\{a, b\} = \{-38, -2\} \text{ yields } c = -21;$$
$$\{a, b\} = \{-12, -8\} \text{ yields } c = -1.$$

Since a and b are interchangeable, each of these cases leads to two solutions, for a total of 12.

29. **(B)** Suppose $1 = x_1 + x_2 + \cdots + x_n$ where $x_1, x_2, \ldots, x_n$ are special and $n \leq 9$. For $k = 1, 2, 3, \ldots$, let a_k be the number of elements of $\{x_1, x_2, \ldots, x_n\}$ whose kth decimal digit is 7. Then

$$1 = \frac{7a_1}{10} + \frac{7a_2}{10^2} + \frac{7a_3}{10^3} + \cdots,$$

which yields

$$\frac{1}{7} = 0.\overline{142857} = \frac{a_1}{10} + \frac{a_2}{10^2} + \frac{a_3}{10^3} + \cdots.$$

Hence $a_1 = 1, a_2 = 4, a_3 = 2, a_4 = 8$, etc. In particular, this implies that $n \geq 8$. On the other hand, $x_1 = 0.\overline{700}$, $x_2 = x_3 = 0.\overline{07}$, $x_4 = x_5 = 0.\overline{077777}$, and $x_6 = x_7 = x_8 = 0.\overline{000777}$ are eight special numbers whose sum is

$$\frac{700700 + 2(70707) + 2(77777) + 3(777)}{999999} = 1.$$

Thus the smallest n is eight.

30. **(C)** In order that $D(n) = 2$, the binary representation of n must consist of a block of 1's followed by a block of 0's followed by a block of 1's. Among the integers n with d-digit binary representations, how many are there for which $D(n) = 2$? If the 0's block consists of just one 0, there are $d - 2$ possible locations for the 0. If the block consists of multiple 0's, then there are $\binom{d-2}{2}$ such blocks, since only the first and last places for the 0's need to be identified. Thus there are $(d-2) + \frac{1}{2}(d-2)(d-3) = \frac{1}{2}(d-2)(d-1)$ values of n with d binary digits such that $D(n) = 2$. The binary representation of 97 has seven digits, so all the 3-, 4-, 5-, and 6-digit binary integers are less than 97. (We need not consider the 1- and 2-digit binary integers.) The sum of the values of $\frac{1}{2}(d-2)(d-1)$ for $d = 3, 4, 5,$ and 6 is 20. We must also consider the 7-digit binary integers less than or equal to $1100001_2 = 97$. If the initial block of 1's contains three or more 1's, then the number would be greater than 97; by inspection, if there are one or two 1's in the initial 1's block, there are respectively five or one acceptable configurations of the 0's block. It follows that the number of solutions of $D(n) = 2$ within the required range is $20 + 5 + 1 = 26$.

OR

Note that $D(n) = 2$ holds exactly when the binary representation of n consists of an initial block of 1's, followed by a block of 0's, and then a final block of 1's. The number of nonnegative integers $n \leq 2^7 - 1 = 127$ for which $D(n) = 2$ is thus $\binom{7}{3} = 35$, since for each n, the corresponding binary representation is given by selecting the position of the leftmost bit in each of the three blocks. If $98 \leq n \leq 127$, the binary representation of n is either (a) $110XXXX_2$ or (b) $111XXXX_2$. Consider those n's for which $D(n) = 2$. By the same argument as above, there are three of type (a), namely $1101111_2 = 111$, $1100111_2 = 103$, and $1100011_2 = 99$. There are $\binom{4}{2} = 6$ of type (b). It follows that the number of solutions of $D(n) = 2$ for which $1 \leq n \leq 97$ is $35 - (3 + 6) = 26$.

49th AHSME solutions, 1998

1. **(E)** Only the rectangle that goes in position II must match on both vertical sides. Since rectangle D is the only one for which these matches exist, it must be the one that goes in position II. Hence the rectangle that goes in position I must be E.

	2			5			1	
9	E	7	7	D	4	4	A	6
	0			8			9	
	0			8				
1	B	3	3	C	5			
	6			2				

2. **(E)** We need to make the numerator large while making the denominator small. The smallest the denominator can be is $0 + 1 = 1$. The largest the numerator can be is $9 + 8 = 17$. The fraction $17/1$ is an integer, so $A + B = 17$.

3. **(D)** The subtraction problem posed is equivalent to the addition problem

$$\begin{array}{r} 4\ 8\ b \\ +\ c\ 7\ 3 \\ \hline 7\ a\ 2 \end{array}$$

   which is easier to solve. Since $b + 3 = 12$, b must be 9. Since $1 + 8 + 7$ has units digit a, a must be 6. Because $1 + 4 + c = 7$, $c = 2$. Hence $a + b + c = 6 + 9 + 2 = 17$.

4. **(E)** Notice that the operation has the property that, for any $r, a, b,$ and c,
$$[ra, rb, rc] = \frac{ra + rb}{rc} = [a, b, c].$$
Thus all three of the expressions $[60, 30, 90], [2, 1, 3],$ and $[10, 5, 15]$ have the same value, which is 1. So $[[60, 30, 90], [2, 1, 3], [10, 5, 15]] = [1, 1, 1] = 2.$

5. **(C)** Factor the left side of the given equation:
$$2^{1998} - 2^{1997} - 2^{1996} + 2^{1995} = (2^3 - 2^2 - 2 + 1)2^{1995} = 3 \cdot 2^{1995} = k \cdot 2^{1995},$$
so $k = 3$.

6. **(C)** The number 1998 has prime factorization $2 \cdot 3^3 \cdot 37$. It has eight factor-pairs: $1 \times 1998 = 2 \times 999 = 3 \times 666 = 6 \times 333 = 9 \times 222 = 18 \times 111 = 27 \times 74 = 37 \times 54 = 1998$. Among these, the smallest difference is $54 - 37 = 17$.

7. **(D)** $\sqrt[3]{N\sqrt[3]{N\sqrt[3]{N}}} = \sqrt[3]{N\sqrt[3]{N \cdot N^{\frac{1}{3}}}} = \sqrt[3]{N\sqrt[3]{N^{\frac{4}{3}}}} =$
$\sqrt[3]{N \cdot N^{\frac{4}{9}}} = \sqrt[3]{N^{\frac{13}{9}}} = N^{\frac{13}{27}}.$

OR

$\sqrt[3]{N\sqrt[3]{N\sqrt[3]{N}}} = \left(N\left(N(N)^{\frac{1}{3}}\right)^{\frac{1}{3}}\right)^{\frac{1}{3}} = \left(N\left(N^{\frac{1}{3}} \cdot N^{\frac{1}{9}}\right)\right)^{\frac{1}{3}} =$
$N^{\frac{1}{3}} \cdot N^{\frac{1}{9}} \cdot N^{\frac{1}{27}} = N^{\frac{13}{27}}.$

8. **(D)** The area of each trapezoid is $1/3$, so $\frac{1}{2} \cdot \frac{1}{2}(x + \frac{1}{2}) = \frac{1}{3}$. Simplifying yields $x + \frac{1}{2} = \frac{4}{3}$, and it follows that $x = 5/6$.

OR

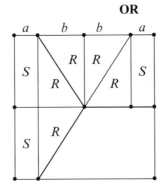

$2S + 2R = S + 3R$
$\therefore S = R$
$\therefore b = 2a$
$a + b + b + a = 1$
$\therefore 2b + a = 5/6$

9. **(D)** Let N be the number of people in the audience. Then $0.2N$ people heard 60 minutes, $0.1N$ heard 0 minutes, $0.35N$ heard 20 minutes, and $0.35N$ heard 40 minutes. In total, the N people heard $60(0.2N) + 0(0.1N) + 20(0.35N) + 40(0.35N) = 12N + 0 + 7N + 14N = 33N$ minutes, so they heard an average of 33 minutes each.

10. **(A)** Let x and y denote the dimensions of the four congruent rectangles. Then $2x + 2y = 14$, so $x + y = 7$. The area of the large square is $(x+y)^2 = 7^2 = 49$.

11. **(D)** The four vertices determine six possible diameters, namely, the four sides and two diagonals. However, the two diagonals are diameters of the same circle. Thus there are five circles.

12. **(A)** Note that $N = 7^{5^{3^{2^{11}}}}$, which has only 7 as a prime factor.

13. **(E)** Factor 144 into primes, $144 = 2^4 \cdot 3^2$, and notice that there are at most two 6's and no 5's among the numbers rolled. If there are no 6's, then there must be two 3's since these are the only values that can contribute 3 to the prime factorization. In this case the four 2's in the factorization must be the result of two 4's in the roll. Hence the sum $3 + 3 + 4 + 4 = 14$ is a possible value for the sum. Next consider the case with just one 6. Then there must be one 3, and the three remaining 2's must be the result of a 4 and a 2. Thus, the sum $6 + 3 + 4 + 2 = 15$ is also possible. Finally, if there are two 6's, then there must also be two 2's or a 4 and a 1, with sums of $6 + 6 + 2 + 2 = 16$ and $6 + 6 + 4 + 1 = 17$. Hence 18 is the only sum not possible.

OR

Since 5 does not divide 144 and $6^3 > 144$, there can be no 5's and at most two 6's. Thus the only ways the four dice can have a sum of 18 are: $4, 4, 4, 6$; $2, 4, 6, 6$; and $3, 3, 6, 6$. Since none of these products is 144, the answer is (E).

14. **(A)** Because the parabola has x-intercepts of opposite sign and the y-coordinate of the vertex is negative, a must be positive, and c, which is the y-intercept, must be negative. The vertex has x-coordinate $-b/2a = 4 > 0$, so b must be negative.

15. **(C)** The regular hexagon can be partitioned into six equilateral triangles, each with area one-sixth of the original triangle. Since the original equilateral triangle is similar to each of these, and the ratio of the areas is 6, it follows that the ratio of the sides is $\sqrt{6}$.

16. **(B)** The area of the shaded region is

$$\frac{\pi}{2}\left(\left(\frac{a+b}{2}\right)^2 + \left(\frac{a}{2}\right)^2 - \left(\frac{b}{2}\right)^2\right) = \frac{\pi}{2}\frac{a+b}{2}\left(\frac{a+b}{2} + \frac{a-b}{2}\right)$$

$$= \frac{\pi(a+b)a}{4}$$

and the area of the unshaded region is

$$\frac{\pi}{2}\left(\left(\frac{a+b}{2}\right)^2 - \left(\frac{a}{2}\right)^2 + \left(\frac{b}{2}\right)^2\right) = \frac{\pi}{2}\frac{a+b}{2}\left(\frac{a+b}{2} + \frac{b-a}{2}\right)$$

$$= \frac{\pi(a+b)b}{4}.$$

Their ratio is a/b.

17. **(E)** Note that $f(x) = f(x+0) = x + f(0) = x + 2$ for any real number x. Hence $f(1998) = 2000$. The function defined by $f(x) = x + 2$ has both properties: $f(0) = 2$ and $f(x+y) = x + y + 2 = x + (y+2) = x + f(y)$.

OR

Note that

$$2 = f(0) = f(-1998 + 1998) = -1998 + f(1998).$$

Hence $f(1998) = 2000$.

18. **(A)** Suppose the sphere has radius r. We can write the volumes of the three solids as functions of r as follows:

$$\text{Volume of cone} = A = \frac{1}{3}\pi r^2(2r) = \frac{2}{3}\pi r^3,$$

$$\text{Volume of cylinder} = M = \pi r^2(2r) = 2\pi r^3, \text{ and}$$

$$\text{Volume of sphere} = C = \frac{4}{3}\pi r^3.$$

Thus, $A - M + C = 0$. **Note.** The AMC logo is designed to show this classical result of Archimedes.

19. **(C)** The area of the triangle is (base)(height)$\div 2 = (5-(-5)) \cdot |5\sin\theta| \div 2 = 25|\sin\theta|$. There are four values of θ between 0 and 2π for which $|\sin\theta| = 0.4$, and each value corresponds to a distinct triangle with area 10.

OR

The vertex $(5\cos\theta, 5\sin\theta)$ lies on a circle of diameter 10 centered at the origin. In order that the triangle have area 10, the altitude from that vertex must be two. There are four points on the circle that are two units from the x-axis.

20. **(C)** There are eight ordered triples of numbers satisfying the conditions: $(1,2,10)$, $(1,3,9)$, $(1,4,8)$, $(1,5,7)$, $(2,3,8)$, $(2,4,7)$, $(2,5,6)$, and $(3,4,6)$. Because Casey's card gives Casey insufficient information, Casey must have seen a 1 or a 2. Next, Tracy must not have seen a 6, 9, or 10, since each of these would enable Tracy to determine the other two cards. Finally, if Stacy had seen a 3 or a 5 on the middle card, Stacy would have been able to determine the other two cards. The only number left is 4, which leaves open the two possible triples $(1,4,8)$ and $(2,4,7)$.

21. **(C)** Let r be Sunny's rate. Thus h/r and $(h+d)/r$ are the times it takes Sunny to cover h meters and $h+d$ meters, respectively. Because Windy covers only $h-d$ meters while Sunny is covering h meters, it follows that Windy's rate is $(h-d)r/h$. While Sunny runs $h+d$ meters, the number of meters Windy runs is $(h-d)r/h \cdot (h+d)/r = (h^2-d^2)/h = h - d^2/h$. Sunny's victory margin over Windy is d^2/h.

OR

Assume $h = 100$, $d = 10$, and the first race ends in 10 seconds. Then Sunny runs 10 m/s and Windy runs 9 m/s. In the second 100-meter race, Sunny runs 110 meters in 11 seconds. Windy covers only $11(9) = 99$ meters in this time, so Sunny finishes 1 meter ahead. Plugging the values for h and d into the selections, only answer A yields a result of 1.

22. **(C)** Express each term using a base-10 logarithm, and note that the sum equals

$$\log 2/\log 100! + \log 3/\log 100! + \cdots + \log 100/\log 100! =$$

$$\log 100!/\log 100! = 1.$$

OR

Since $1/\log_k 100!$ equals $\log_{100!} k$ for all positive integers k, the expression equals $\log_{100!}(2 \cdot 3 \cdots 100) = \log_{100!} 100! = 1$.

23. **(D)** Complete the squares in the two equations to bring them to the form

$$(x-6)^2 + (y-3)^2 = 7^2 \quad \text{and} \quad (x-2)^2 + (y-6)^2 = k+40.$$

The graphs of these equations are circles. The first circle has radius 7, and the distance between the centers of the circles is 5. In order for the circles to have a point in common, therefore, the radius of the second circle must be at least 2 and at most 12. It follows that $2^2 \le k+40 \le 12^2$, or $-36 \le k \le 104$. Thus $b - a = 140$.

24. **(C)** There are 10,000 ways to write the last four digits $d_4d_5d_6d_7$, and among these there are $10000 - 10 = 9990$ for which not all the digits are the same. For each of these, there are exactly two ways to adjoin the three digits $d_1d_2d_3$ to obtain a memorable number. There are ten memorable numbers for which the last four digits are the same, for a total of $2 \cdot 9990 + 10 = 19990$.

OR

Let A denote the set of telephone numbers for which $d_1d_2d_3$ and $d_4d_5d_6$ are identical and B the set for which $d_1d_2d_3$ is the same as $d_5d_6d_7$. A number $d_1d_2d_3\text{-}d_4d_5d_6d_7$ belongs to $A \cap B$ if and only if $d_1 = d_4 = d_5 = d_2 = d_6 = d_3 = d_7$. Hence, $n(A \cap B) = 10$. Thus, by the *Inclusion-Exclusion Principle*, $n(A \cup B) =$

$$n(A) + n(B) - n(A \cap B) = 10^3 \cdot 1 \cdot 10 + 10^3 \cdot 10 \cdot 1 - 10 = 19990.$$

25. **(B)** The crease in the paper is the perpendicular bisector of the segment that joins $(0, 2)$ to $(4, 0)$. Thus the crease contains the midpoint $(2, 1)$ and has slope 2, so the equation $y = 2x - 3$ describes it. The segment joining $(7, 3)$ and (m, n) must have slope $-1/2$, and its midpoint $(7+m)/2, (3+n)/2$ must also satisfy the equation $y = 2x - 3$. It follows that

$$-\frac{1}{2} = \frac{n-3}{m-7} \quad \text{and} \quad \frac{3+n}{2} = 2 \cdot \frac{7+m}{2} - 3, \text{ so}$$

$$2n + m = 13 \quad \text{and} \quad n - 2m = 5.$$

Solve these equations simultaneously to find that $m = 3/5$ and $n = 31/5$, so that $m + n = 34/5 = 6.8$.

OR

As shown above, the crease is described by the equation $y = 2x - 3$. Therefore, the slope of the line through (m, n) and $(7, 3)$ is $-1/2$, so the points on the line can be described parametrically by $(x, y) = (7 - 2t, 3 + t)$. The intersection of this line with the crease $y = 2x - 3$ is found by solving $3 + t = 2(7 - 2t) - 3$. This yields the parameter value $t = 8/5$. Since $t = 8/5$ determines the point on the crease, use $t = 2(8/5)$ to find the coordinates $m = 7 - 2(16/5) = 3/5$ and $n = 3 + (16/5) = 31/5$.

26. **(B)** Extend $\overline{DA}$ through A and $\overline{CB}$ through B and denote the intersection by E. Triangle ABE is a 30°-60°-90° triangle with $AB = 13$, so $AE = 26$. Triangle CDE is also a 30°-60°-90° triangle, from which it follows that $CD = (46 + 26)/\sqrt{3} = 24\sqrt{3}$. Now apply the *Pythagorean Theorem* to triangle CDA to find that $AC = \sqrt{46^2 + (24\sqrt{3})^2} = 62$.

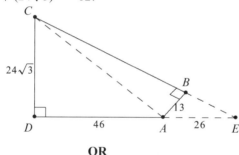

OR

Since the opposite angles sum to a straight angle, the quadrilateral is cyclic, and AC is the diameter of the circumscribed circle. Thus AC is the diameter of the circumcircle of triangle ABD. By the *Extended Law of Sines*,
$$AC = \frac{BD}{\sin 120°} = \frac{BD}{\sqrt{3}/2}.$$

We determine BD by the *Law of Cosines*:

$$BD^2 = 13^2 + 46^2 + 2 \cdot 13 \cdot 46 \div 2 = 2883 = 3 \cdot 31^2, \text{ so } BD = 31\sqrt{3}.$$

Hence $AC = 62$.

27. **(E)** After step one, twenty $3 \times 3 \times 3$ cubes remain, eight of which are corner cubes and twelve of which are edge cubes. At this stage each $3 \times 3 \times 3$ corner cube contributes 27 units of area and each $3 \times 3 \times 3$ edge cube contributes 36 units of area. The second stage of the tunneling process takes away 3 units of area from each of the eight $3 \times 3 \times 3$ corner cubes (1 for each exposed surface), but adds 24 units to the area (4 units for each of the six 1×1 center facial cubes removed). The twelve $3 \times 3 \times 3$ edge cubes each lose 4 units but gain 24 units. Therefore, the total surface area of the figure is

$$8 \cdot (27 - 3 + 24) + 12 \cdot (36 - 4 + 24) = 384 + 672 = 1056.$$

28. **(B)** Let E denote the point on $\overline{BC}$ for which $\overline{AE}$ bisects $\angle CAD$. Because the answer is not changed by a similarity transformation, we may assume that $AC = 2\sqrt{5}$ and $AD = 3\sqrt{5}$. Apply the *Pythagorean Theorem* to triangle ACD to obtain $CD = 5$, then apply the *Angle Bisector Theorem* to triangle CAD to obtain $CE = 2$ and $ED = 3$. Let $x = DB$. Apply the Pythagorean Theorem to triangle ACE to obtain $AE = \sqrt{24}$, then apply the Angle Bisector Theorem to triangle EAB to obtain $AB = (x/3)\sqrt{24}$. Now apply the Pythagorean Theorem to triangle ABC to get

$$(2\sqrt{5})^2 + (x+5)^2 = \left(\frac{x}{3}\sqrt{24}\right)^2,$$

from which it follows that $x = 9$. Hence $BD/DC = 9/5$, and $m + n = 14$.

OR

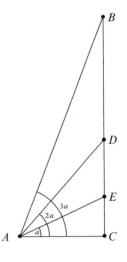

Denote by a the measure of angle CAE. Let $AC = 2u$, and $AD = 3u$. It follows that $CD = \sqrt{5}u$. We may assume $BD = \sqrt{5}$. (Otherwise, we could simply modify the triangle with a similarity transformation.) Hence, the ratio CD/BD we seek is just u. Since $\cos 2a = 2/3$, we have $\sin a = 1/\sqrt{6}$. Applying the *Law of Sines* in triangle ABD yields

$$\frac{\sin D}{AB} = \frac{\sin a}{\sqrt{5}} = \frac{2/3}{\sqrt{(2u)^2 + (\sqrt{5}(1+u))^2}}$$

$$= \frac{1/\sqrt{6}}{\sqrt{5}}.$$

Solve this for u to get

$$2\sqrt{5}\sqrt{6} = 3\sqrt{4u^2 + 5(1 + 2u + u^2)}$$
$$120 = 9(9u^2 + 10u + 5)$$
$$0 = 27u^2 + 30u - 25$$
$$0 = (9u - 5)(3u + 5)$$

so $u = 5/9$ and $m + n = 14$.

OR

Again, let $a = \angle CAE$. We are given that $\cos 2a = 2/3$ and we wish to compute

$$\frac{CD}{BD} = \frac{AC \tan 2a}{AC(\tan 3a - \tan 2a)} = \left(\frac{\tan 3a}{\tan 2a} - 1\right)^{-1}.$$

Let $y = \tan a$. Trigonometric identities yield (upon simplification)

$$\left(\frac{\tan 3a}{\tan 2a} - 1\right)^{-1} = \frac{2(1 - 3y^2)}{(1 + y^2)^2} \quad \text{and} \quad \frac{2}{3} = \cos 2a = \frac{1 - y^2}{1 + y^2}.$$

Thus $y^2 = 1/5$ and

$$\frac{CD}{BD} = \frac{2(1 - 3/5)}{(6/5)^2} = \frac{5}{9}.$$

Alternatively, starting with $a = \cos^{-1}(2/3)/2$, electronic calculation yields $\tan(3a)/\tan(2a) = 2.8 = 14/5$, so $CD/BD = 5/9$.

29. **(D)** If a square encloses three collinear lattice points, then it is not hard to see that the square must also enclose at least one additional lattice point. It therefore suffices to consider squares that enclose only the lattice points (0,0), (0,1), and (1,0). If a square had two adjacent sides, neither of which contained a lattice point, then the square could be enlarged slightly by moving those sides parallel to themselves. To be largest, therefore, a square must contain a lattice point on at least

two nonadjacent sides. The desired square will thus have parallel sides that contain $(1,1)$ and at least one of $(-1,0)$ and $(0,-1)$. The size of the square is determined by the separation between two parallel sides. Because the distance between parallel lines through $(1,1)$ and $(0,-1)$ can be no larger than $\sqrt{5}$, the largest conceivable area for the square is 5. To see that this is in fact possible, draw the lines of slope 2 through $(-1,0)$ and $(1,-1)$, and the lines of slope $-1/2$ through $(1,1)$ and $(0,-1)$. These four lines can be described by the equations $y = 2x + 2$, $y = 2x - 3$, $2y + x = 3$, and $2y + x = -2$, respectively. They intersect to form a square whose area is 5, and whose vertices are $(-1/5, 8/5)$, $(9/5, 3/5)$, $(4/5, -7/5)$, and $(-6/5, -2/5)$. There are only three lattice points inside this square.

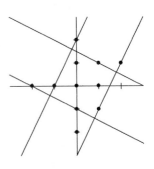

30. **(E)** Factor a_n as a product of prime powers:

$$a_n = n(n+1)(n+2)\cdots(n+9) = 2^{e_1} 3^{e_2} 5^{e_3} \cdots.$$

Among the ten factors $n, n+1, \ldots, n+9$, five are even and their product can be written $2^5 m(m+1)(m+2)(m+3)(m+4)$. If m is even, then $m(m+2)(m+4)$ is divisible by 16 and thus $e_1 \geq 9$. If m is odd, then $e_1 \geq 8$. If $e_1 > e_3$, then the rightmost nonzero digit of a_n is even. If $e_1 \leq e_3$, then the rightmost nonzero digit of a_n is odd. Hence we seek the smallest n for which $e_3 \geq e_1$. Among the ten numbers $n, n+1, \ldots, n+9$, two are divisible by 5 and at most one of these is divisible by 25. Hence $e_3 \geq 8$ if and only if one of $n, n+1, \ldots, n+9$ is divisible by 5^7. The smallest n for which a_n satisfies $e_3 \geq 8$ is thus $n = 5^7 - 9$, but in this case the product of the five even numbers among $n, n+1, \ldots, n+9$ is $2^5 m(m+1)(m+2)(m+3)(m+4)$ where m is even, namely $(5^7 - 9)/2 = 39058$. As noted earlier, this gives $e_1 \geq 9$. For $n = 5^7 - 8 = 78117$, the product of the five even numbers among $n, n+1, \ldots, n+9$ is $2^5 m(m+1)(m+2)(m+3)(m+4)$ with $m = 39059$. Note that in this case $e_1 = 8$. Indeed, $39059 + 1$ is divisible by 4 but not by 8, and $39059 + 3$ is divisible by 2 but not by 4. Compute the rightmost nonzero digit as follows. The odd numbers among $n, n+1, \ldots, n+9$ are $78117, 78119, 78121, 78123, 78125 = 5^7$ and the product of the even numbers 78118, 78120, 78122, 78124, 78126 is $2^5 \cdot 39059 \cdot 39060 \cdot 39061 \cdot 39062 \cdot 39063 = 2^5 \cdot 3905\underline{9} \cdot (2^2 \cdot 5 \cdot 195\underline{3}) \cdot 3906\underline{1} \cdot (2 \cdot 1953\underline{1}) \cdot 3906\underline{3}$. (For convenience, we have underlined the needed unit digits.) Having written $n(n+1)\cdots(n+9)$ as $2^8 5^8$ times a product of odd factors not divisible by 5, we determine the rightmost nonzero digit by multiplying the units digits of these factors. It follows that, for $n = 5^7 - 8$, the rightmost nonzero digit of a_n is the units digit of $7 \cdot 9 \cdot 1 \cdot 3 \cdot 9 \cdot 3 \cdot 1 \cdot 1 \cdot 3 = (9 \cdot 9) \cdot (7 \cdot 3) \cdot (3 \cdot 3)$, namely 9.

50th AHSME solutions, 1999

1. **(E)** Pairing the first two terms, the next two terms, etc. yields
$$1 - 2 + 3 - 4 + \cdots - 98 + 99 =$$
$$(1 - 2) + (3 - 4) + \cdots + (97 - 98) + 99 =$$
$$-1 - 1 - 1 - \cdots - 1 + 99 = 50,$$
since there are 49 of the -1's.

OR

$$1 - 2 + 3 - 4 + \cdots - 98 + 99 =$$
$$1 + [(-2 + 3) + (-4 + 5) + \cdots + (-98 + 99)] =$$
$$1 + [1 + 1 + \cdots + 1] = 1 + 49 = 50.$$

2. **(A)** Triangles with side lengths of 1, 1, 1 and 2, 2, 2 are equilateral and not congruent, so (A) is false. Statement (B) is true since all triangles are convex. Statements (C) and (E) are true since each interior angle of an equilateral triangle measures 60°. Furthermore, all three sides of an equilateral triangle have the same length, so (D) is also true.

3. **(E)** The desired number is the arithmetic average or mean:
$$\frac{1}{2}\left(\frac{1}{8} + \frac{1}{10}\right) = \frac{1}{2} \cdot \frac{18}{80} = \frac{9}{80}.$$

4. **(A)** A number one less than a multiple of 5 is has a units digit of 4 or 9. A number whose units digit is 4 cannot be one greater than a multiple of 4. Thus, it is sufficient to examine the numbers of the form

$10d + 9$ where d is one of the ten digits. Of these, only $9, 29, 49, 69$ and 89 are one greater than a multiple of 4. Among these, only 29 and 89 are prime and their sum is 118.

5. **(C)** If the suggested retail price was P, then the marked price was $0.7P$. Half of this is $0.35P$, so Alice paid 35% of the suggested retail price.

6. **(D)** Note that

$$2^{1999} \cdot 5^{2001} = 2^{1999} \cdot 5^{1999} \cdot 5^2 = 10^{1999} \cdot 25 = 25\overbrace{0\ldots0}^{1999 \text{ zeros}}.$$

Hence the sum of the digits is 7.

7. **(B)** The sum of the angles in a convex hexagon is $720°$ and each angle must be less than $180°$. If four of the angles are acute, then their sum would be less than $360°$, and therefore at least one of the two remaining angles would be greater than $180°$, a contradiction. Thus there can be at most three acute angles. The hexagon shown has three acute angles, A, C, and E.

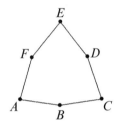

OR

The result holds for *any* convex n-gon. The sum of the exterior angles of a convex n-gon is $360°$. Hence at most three of these angles can be obtuse, for otherwise the sum would exceed $360°$. Thus the largest number of acute angles in any convex n-gon is three.

8. **(D)** Let w and $2w$ denote the ages of Walter and his grandmother, respectively, at the end of 1994. Then their respective years of birth are $1994-w$ and $1994-2w$. Hence $(1994-w)+(1994-2w) = 3838$, and it follows that $w = 50$ and Walter's age at the end of 1999 will be 55.

9. **(D)** The next palindromes after 29792 are $29892, 29992, 30003,$ and 30103. The difference $30103 - 29792 = 311$ is too far to drive in three hours without exceeding the speed limit of 75 miles per hour. Ashley could have driven $30003 - 29792 = 211$ miles during the three hours for an average speed of $70\frac{1}{3}$ miles per hour.

10. **(C)** Since both I and III cannot be false, the digit must be 1 or 3. So either I or III is the false statement. Thus II and IV must be true and (C) is necessarily correct. For the same reason, (E) must be incorrect. If the digit is 1, (B) and (D) are incorrect, and if the digit is 3, (A) is incorrect.

11. **(A)** The locker labelling requires $137.94/0.02 = 6897$ digits. Lockers 1 through 9 require 9 digits, lockers 10 through 99 require $2 \cdot 90 = 180$ digits, and lockers 100 through 999 require $3 \cdot 900 = 2700$ digits. Hence the remaining lockers require $6897 - 2700 - 180 - 9 = 4008$ digits, so there must be $4008/4 = 1002$ more lockers, each using four digits. In all, there are $1002 + 999 = 2001$ student lockers.

12. **(C)** The x-coordinates of the intersection points are precisely the zeros of the polynomial $p(x) - q(x)$. This polynomial has degree at most three, so it has at most three zeros. Hence, the graphs of the fourth degree polynomial functions intersect at most three times. Finding an example to show that three intersection points can be achieved is left to the reader.

13. **(C)** Since $a_{n+1} = \sqrt[3]{99} \cdot a_n$ for all $n \geq 1$, it follows that $a_1, a_2, a_3, \ldots$ is a geometric sequence whose first term is 1 and whose common ratio is $r = \sqrt[3]{99}$. Thus
$$a_{100} = a_1 \cdot r^{100-1} = \left(\sqrt[3]{99}\right)^{99} = 99^{33}.$$

14. **(A)** Tina and Alina each sang either 5 or 6 times. If N denotes the number of songs sung by trios, then $3N = 4 + 5 + 5 + 7 = 21$ or $3N = 4 + 5 + 6 + 7 = 22$ or $3N = 4 + 6 + 6 + 7 = 23$. Since the girls sang as trios, the total must be a multiple of 3. Only 21 qualifies. Therefore, $N = 21/3 = 7$ is the number of songs the trios sang. **Challenge.** Devise a schedule for the four girls so that each one sings the required number of songs.

15. **(E)** From the identity $1 + \tan^2 x = \sec^2 x$ it follows that $1 = \sec^2 x - \tan^2 x = (\sec x - \tan x)(\sec x + \tan x) = 2(\sec x + \tan x)$, so $\sec x + \tan x = 0.5$.

OR

The given relation can be written as $(1 - \sin x) \div \cos x = 2$. Squaring both sides yields $(1 - \sin x)^2 \div (1 - \sin^2 x) = 4$, hence $(1 - \sin x) \div (1 + \sin x) = 4$. It follows that $\sin x = 3/5$ and that $\cos x = $

$(1 - \sin x)/2 = (1 - (-3/5))/2 = 4/5$. Thus $\sec x + \tan x = 5/4 - 3/4 = 0.5$.

16. **(C)** Let E be the intersection of the diagonals of a rhombus $ABCD$ satisfying the conditions of the problem. Because these diagonals are perpendicular and bisect each other, $\triangle ABE$ is a right triangle with sides 5, 12, and 13 and area 30. Therefore the altitude drawn to side AB is $60/13$, which is the radius of the inscribed circle centered at E.

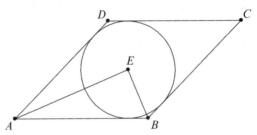

17. **(C)** From the hypothesis, $P(19) = 99$ and $P(99) = 19$. Let

$$P(x) = (x - 19)(x - 99)Q(x) + ax + b,$$

where a and b are constants and $Q(x)$ is a polynomial. Then

$$99 = P(19) = 19a + b \text{ and } 19 = P(99) = 99a + b.$$

It follows that $99a - 19a = 19 - 99$, hence $a = -1$ and $b = 99 + 19 = 118$. Thus the remainder is $-x + 118$.

18. **(E)** Note that the range of $\log x$ on the interval $(0, 1)$ is the set of all negative numbers, infinitely many of which are zeros of the cosine function. In fact, since $\cos(x) = 0$ for all x of the form $\pi/2 \pm n\pi$,

$$f(10^{\frac{\pi}{2} - n\pi}) = \cos(\log(10^{\frac{\pi}{2} - n\pi}))$$
$$= \cos\left(\frac{\pi}{2} - n\pi\right)$$
$$= 0$$

for all positive integers n.

19. **(C)** Let $DC = m$ and $AD = n$. By the *Pythagorean Theorem*, $AB^2 = AD^2 + DB^2$. Hence $(m + n)^2 = n^2 + 57$, which yields $m(m + 2n) = 57$. Since m and n are positive integers, the only possibilities are $m = 1$, $n = 28$ and $m = 3$, $n = 8$. The second of these gives the least possible value of $AC = m + n$, namely 11.

20. **(E)** For $n \geq 3$,
$$a_n = \frac{a_1 + a_2 + \cdots + a_{n-1}}{n-1}.$$
Thus $(n-1)a_n = a_1 + a_2 + \cdots + a_{n-1}$. It follows that
$$a_{n+1} = \frac{a_1 + a_2 + \cdots + a_{n-1} + a_n}{n} = \frac{(n-1) \cdot a_n + a_n}{n} = a_n,$$
for $n \geq 3$. Since $a_9 = 99$ and $a_1 = 19$, it follows that
$$99 = a_3 = \frac{19 + a_2}{2},$$
and hence that $a_2 = 179$. (The sequence is $19, 179, 99, 99, \ldots$.)

21. **(B)** Since $20^2 + 21^2 = 29^2$, the converse of the *Pythagorean Theorem* applies, so the triangle has a right angle. Thus its hypotenuse is a diameter of the circle, so the region with area C is a semicircle and is congruent to the semicircle formed by the other three regions. The area of the triangle is 210, hence $A + B + 210 = C$. To see that the other options are incorrect, note that **(A)** $A + B < A + B + 210 = C$; **(C)** $A^2 + B^2 < (A+B)^2 < (A+B+210)^2 = C^2$; **(D)** $20A + 21B < 29A + 29B < 29(A + B + 210) = 29C$; and **(E)** $\frac{1}{A^2} + \frac{1}{B^2} > \frac{1}{A^2} > \frac{1}{C^2}$.

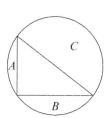

22. **(C)** The first graph is an inverted 'V-shaped' right angle with vertex at (a, b) and the second is a V-shaped right angle with vertex at (c, d). Thus (a, b), $(2, 5)$, (c, d), and $(8, 3)$ are consecutive vertices of a rectangle. The diagonals of this rectangle meet at their common midpoint, so the x-coordinate of this midpoint is $(2 + 8)/2 = (a + c)/2$. Thus $a + c = 10$.

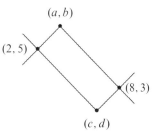

OR

Use the given information to obtain the equations $5 = -|2 - a| + b$, $5 = |2-c|+d$, $3 = -|8-a|+b$, and $3 = |8-c|+d$. Subtract the third from the first to eliminate b and subtract the fourth from the second to eliminate d. The two resulting equations $|8 - a| - |2 - a| = 2$ and

$|2-c|-|8-c| = 2$ can be solved for a and c. To solve the former, first consider all $a \leq 2$, for which the equation reduces to $8-a-(2-a) = 2$, which has no solutions. Then consider all a in the interval $2 \leq a \leq 8$, for which the equation reduces to $8 - a - (a - 2) = 2$, which yields $a = 4$. Finally, consider all $a \geq 8$, for which the equation reduces to $a - 8 - (a - 2) = 2$, which has no solutions. The other equation can be solved similarly to show that $c = 6$. Thus $a + c = 10$.

23. **(E)** Extend $\overline{FA}$ and $\overline{CB}$ to meet at X, $\overline{BC}$ and $\overline{ED}$ to meet at Y, and $\overline{DE}$ and $\overline{AF}$ to meet at Z. The interior angles of the hexagon are $120°$. Thus the triangles XYZ, ABX, CDY, and EFZ are equilateral. Since $AB = 1$, $BX = 1$. Since $CD = 2$, $CY = 2$. Thus $XY = 7$ and $YZ = 7$. Since $YD = 2$ and $DE = 4$, $EZ = 1$. The area of the hexagon can be found by subtracting the areas of the three small triangles from the area of the large triangle:

$$7^2 \left(\frac{\sqrt{3}}{4}\right) - 1^2 \left(\frac{\sqrt{3}}{4}\right) - 2^2 \left(\frac{\sqrt{3}}{4}\right) - 1^2 \left(\frac{\sqrt{3}}{4}\right) = \frac{43\sqrt{3}}{4}.$$

24. **(B)** Any four of the six given points determine a unique convex quadrilateral, so there are exactly $\binom{6}{4} = 15$ favorable outcomes when the chords are selected randomly. Since there are $\binom{6}{2} = 15$ chords, there are $\binom{15}{4} = 1365$ ways to pick the four chords. So the desired probability is $15/1365 = 1/91$.

25. **(B)** Multiply both sides of the equation by $7!$ to obtain

$$3600 = 2520a_2 + 840a_3 + 210a_4 + 42a_5 + 7a_6 + a_7.$$

It follows that $3600 - a_7$ is a multiple of 7, which implies that $a_7 = 2$. Thus,

$$\frac{3598}{7} = 514 = 360a_2 + 120a_3 + 30a_4 + 6a_5 + a_6.$$

Reason as above to show that $514 - a_6$ is a multiple of 6, which implies that $a_6 = 4$. Thus, $510/6 = 85 = 60a_2 + 20a_3 + 5a_4 + a_5$. Then it follows that $85 - a_5$ is a multiple of 5, whence $a_5 = 0$. Continue in this fashion to obtain $a_4 = 1, a_3 = 1$, and $a_2 = 1$. Thus the desired sum is $1 + 1 + 1 + 0 + 4 + 2 = 9$.

OR

Note that, if $0 \leq a_i \leq i-1$ for all i, then

$$\frac{a_m}{m!} + \cdots + \frac{a_n}{n!} \leq \frac{m-1}{m!} + \cdots + \frac{n-1}{n!}$$

$$= \left(\frac{1}{(m-1)!} - \frac{1}{m!}\right) + \cdots + \left(\frac{1}{(n-1)!} - \frac{1}{n!}\right)$$

$$= \frac{1}{(m-1)!} - \frac{1}{n!} < \frac{1}{(m-1)!}.$$

Setting $m = k+1$ in this inequality, we see that

$$r_k = \frac{a_k}{k!} + \left(\frac{a_{k+1}}{(k+1)!} + \cdots + \frac{a_n}{n!}\right) < \frac{a_k}{k!} + \frac{1}{k!}.$$

Thus $\frac{a_k}{k!} < r_k < \frac{a_k+1}{k!}$, so that $a_k < k! \cdot r_k < a_k + 1$. It follows that

$$a_k = \lfloor k! \cdot r_k \rfloor.$$

Repeated use of this fact yields

$$\frac{6}{7} = \frac{a_2}{2!} + \frac{a_3}{3!} + \cdots + \frac{a_7}{7!} \Rightarrow a_2 = \lfloor 2! \cdot 6/7 \rfloor = 1$$

$$\frac{6}{7} - \frac{1}{2!} = \frac{5}{14} = \frac{a_3}{3!} + \frac{a_4}{4!} + \cdots + \frac{a_7}{7!} \Rightarrow a_3 = \lfloor 3! \cdot 5/14 \rfloor = 2$$

$$\frac{5}{14} - \frac{2}{3!} = \frac{1}{42} = \frac{a_4}{4!} + \frac{a_5}{5!} + \frac{a_6}{6!} + \frac{a_7}{7!} \Rightarrow a_4 = \lfloor 4! \cdot 1/42 \rfloor = 0$$

$$\frac{1}{42} - \frac{0}{4!} = \frac{1}{42} = \frac{a_5}{5!} + \frac{a_6}{6!} + \frac{a_7}{7!} \Rightarrow a_5 = \lfloor 5! \cdot 1/42 \rfloor = 2$$

$$\frac{1}{42} - \frac{2}{5!} = \frac{1}{140} = \frac{a_6}{6!} + \frac{a_7}{7!} \Rightarrow a_6 = \lfloor 6! \cdot 1/140 \rfloor = 5$$

$$\frac{1}{140} - \frac{5}{6!} = \frac{1}{5040} = \frac{a_7}{7!} \Rightarrow a_7 = 1.$$

26. **(D)** The interior angle of a regular n-gon is $180(1-2/n)$. Let a be the number of sides of the congruent polygons and let b be the number of sides of the third polygon (which could be congruent to the first two polygons). Then

$$2 \cdot 180\left(1 - \frac{2}{a}\right) + 180\left(1 - \frac{2}{b}\right) = 360.$$

Clearing denominators and factoring yields the equation
$$(a-4)(b-2) = 8,$$
whose four positive integral solutions are $(a, b) = (5, 10), (6, 6), (8, 4)$, and $(12, 3)$. These four solutions give rise to polygons with perimeters of 14, 12, 14 and 21, respectively, so the largest possible perimeter is 21.

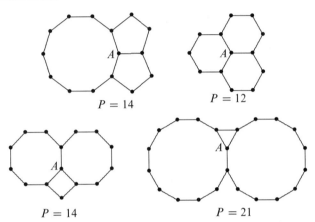

27. **(A)** Square both sides of the equations and add the results to obtain $9(\sin^2 A + \cos^2 A) + 16(\sin^2 B + \cos^2 B) + 24(\sin A \cos B + \sin B \cos A) = 37$. Hence, $24\sin(A + B) = 12$. Thus $\sin C = \sin(180° - A - B) = \sin(A + B) = \frac{1}{2}$, so $\angle C = 30°$ or $\angle C = 150°$. The latter is impossible because it would imply that $A < 30°$ and consequently that $3 \sin A + 4 \cos B < 3 \cdot \frac{1}{2} + 4 < 6$, a contradiction. Therefore $\angle C = 30°$. **Challenge.** Prove that there is a unique such triangle (up to similarity), the one for which $\cos A = (5 - 12\sqrt{3})/37$ and $\cos B = (66 - 3\sqrt{3})/74$.

28. **(E)** Let $a, b,$ and c denote the number of -1's, 1's, and 2's in the sequence, respectively. We need not consider the zeros. Then a, b, c are nonnegative integers satisfying $-a + b + 2c = 19$ and $a + b + 4c = 99$. It follows that $a = 40 - c$ and $b = 59 - 3c$, where $0 \le c \le 19$ (since $b \ge 0$), so
$$x_1^3 + x_2^3 + \cdots + x_n^3 = -a + b + 8c = 19 + 6c.$$
The lower bound is achieved when $c = 0$ ($a = 40, b = 59$). The upper bound is achieved when $c = 19$ ($a = 21, b = 2$). Thus $m = 19$ and $M = 133$, so $M/m = 7$.

29. **(C)** Let A, B, C, and D be the vertices of the tetrahedron. Let O be the center of both the inscribed and circumscribed spheres. Let the inscribed sphere be tangent to the face ABC at the point E, and let its volume be V. Note that the radius of the inscribed sphere is OE and the radius of the circumscribed sphere is OD. Draw $\overline{OA}$, $\overline{OB}$, $\overline{OC}$, and $\overline{OD}$ to obtain four congruent tetrahedra $ABCO$, $ABDO$, $ACDO$, and $BCDO$, each with volume $1/4$ that of the original tetrahedron. Because the two tetrahedra $ABCD$ and $ABCO$ share the same base, $\triangle ABC$, the ratio of the distance from O to face ABC to the distance from D to face ABC is $1/4$; that is, $OD = 3 \cdot OE$. Thus the volume of the circumscribed sphere is $27V$. Extend $\overline{DE}$ to meet the circumscribed sphere at F. Then $DF = 2 \cdot DO = 6 \cdot OE$. Thus $EF = 2 \cdot OE$, so the sphere with diameter $\overline{EF}$ is congruent to the inscribed sphere, and thus has volume V. Similarly each of the other three spheres between the tetrahedron and the circumscribed sphere have volume V. The five congruent small spheres have no volume in common and lie entirely inside the circumscribed sphere, so the ratio $5V/27V$ is the probability that a point in the circumscribed sphere also lies in one of the small spheres. The fraction $5/27$ is closer to 0.2 than it is to any of the other choices.

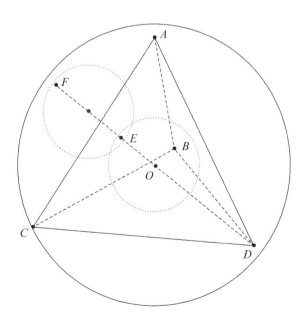

30. **(D)** Let $m+n = s$. Then $m^3 + n^3 + 3mn(m+n) = s^3$. Subtracting the given equation from the latter yields

$$s^3 - 33^3 = 3mns - 99mn.$$

It follows that $(s - 33)(s^2 + 33s + 33^2 - 3mn) = 0$, hence either $s = 33$ or $(m+n)^2 + 33(m+n) + 33^2 - 3mn = 0$. The second equation is equivalent to $(m-n)^2 + (m+33)^2 + (n+33)^2 = 0$, whose only solution, $(-33, -33)$, qualifies. On the other hand, the solutions to $m + n = 33$ satisfying the required conditions are $(0, 33), (1, 32), (2, 31), \ldots, (33, 0)$, of which there are 34. Thus there are 35 solutions altogether.

Sample AMC 10 solutions, 1999

1. **(D)** The desired number is the arithmetic average or mean:
$$\frac{1}{2}\left(\frac{1}{6}+\frac{1}{4}\right) = \frac{1}{2}\cdot\frac{10}{24} = \frac{5}{24}.$$

2. **(D)** If the suggested retail price was P, then the marked price was $0.6P$. Half of this is $0.3P$, which is 70% less than the suggested retail price.

3. **(D)** Let $a, b,$ and c denote the three numbers, with $a < b < c$. Then $(a + b + c) \div 3 = a + 10 = c - 15$. Also, $b = 5$. Thus $a + c + 5 = 3a + 30$, so $c - 2a = 25$ and $c = 25 + 2a$. Therefore, $c - 15 = 25 + 2a - 15 = a + 10$ and it follows that $a = 0$, $b = 5$, and $c = 25$. Thus $a + b + c = 30$.

4. **(D)** The sum of the interior angles of a quadrilateral is 360°. Since an obtuse angle is greater than 90°, the sum of three obtuse angles is greater than 270°. Thus, if a quadrilateral has three obtuse angles, its fourth angle must be less than 90°, and therefore not obtuse. The quadrilateral shown has three obtuse angles.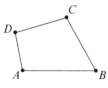

5. **(B)** The terms can be paired
$$(1-2) + (3-4) + (5-6) + \cdots + (199-200) = -1 - 1 - 1 \cdots - 1$$
$$= -100,$$
so the average value is $-100/200 = -0.5$.

6. **(D)** Note that
$$2^{1999} \cdot 5^{2001} = 2^{1999} \cdot 5^{1999} \cdot 5^2 = 10^{1999} \cdot 25 = 25\overbrace{0\ldots 0}^{1999 \text{ zeros}}.$$
Hence the sum of the digits is 7.

7. **(A)** A number one less than a multiple of 5 is has a units digit of 4 or 9. A number whose units digit is 4 cannot be one greater than a multiple of 4. Thus, it is sufficient to examine the numbers of the form $10d + 9$ where d is one of the ten digits. Of these, only 9, 29, 49, 69 and 89 are one greater than a multiple of 4. Among these, only 29 and 89 are prime and their sum is 118.

8. **(D)** The line bisecting the two rectangles must pass through the center of each rectangle, $(-1, 2)$ and $(3, 6)$. Therefore, the slope of the line is

$$(6 - 2)/(3 - (-1)) = 1.$$

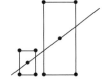

9. **(E)** The value of a $3 \times 3 \times 3$ cube is $(27/8)200 = 27 \cdot 25 = 625$.

10. **(A)** Each $m \times n$ face has $(m-2) \cdot (n-2)$ unit cubes painted on just one face. Thus, there are $2(4-2)(6-2)+2(4-2)(8-2)+2(6-2)(8-2) = 2 \cdot 2 \cdot 4 + 2 \cdot 2 \cdot 6 + 2 \cdot 4 \cdot 6 = 16 + 24 + 48 = 88$ cubes with just one face painted. Note that **B** is the number of cubes with one or two faces painted, **C** is the number of cubes on the surface, **D** is the volume, and **E** is the surface area of the block.

11. **(D)** The sum of the lengths of the four horizontal segments that form the top boundary is 12, and the sum of the lengths of the segments forming the left and right boundaries are 10, so the perimeter is $12 + 12 + 10 + 10 = 44$.

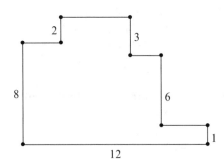

12. **(C)** The number 50 can be written as a sum of squares of distinct digits in several ways: $25 + 25$; $49 + 1$; $25 + 16 + 9$; and $36 + 9 + 4 + 1$. The one with four summands produces the largest integer, 1236. The product of the digits is 36.

13. **(C)** Let w and $2w$ denote the ages of Walter and his grandmother, respectively, at the end of 1994. Then their respective years of birth are $1994-w$ and $1994-2w$. Hence $(1994-w)+(1994-2w) = 3844$, and it follows that $w = 48$ and Walter's age at the end of 1999 will be 53.

14. **(D)** The units digit of the product depends only on the units digits of the factors. The units digit of $2 \cdot 4 \cdot 6 \cdot 8$ is 4, so the problem is reduced to finding the units digit of 4^{10}, which is the same as that of 6^5, which is 6.

15. **(C)** To get an even sum, either all three summands must be even or exactly one must be even. There is one 3-element subset of the first type and $3 \cdot 3 = 9$ subsets of the second type because there are 3 ways to choose the even member and 3 ways to choose the two odd members.

16. **(A)** Let r denote the radius of the circle. A point P is closer to the center of the circle than it is to the boundary precisely when P belongs to the circle of radius $r/2$ with the same center. The probability that this occurs is
$$\frac{\pi(r/2)^2}{\pi r^2} = \frac{1}{4}.$$

17. **(B)** The locker labeling requires $145.86/0.02 = 7293$ digits. Lockers 1 through 9 require 9 digits, lockers 10 through 99 require $2 \cdot 90 = 180$ digits, and lockers 100 through 999 require $3 \cdot 900 = 2700$ digits. Hence the remaining lockers require $7293 - 2700 - 180 - 9 = 4404$ digits, so there must be $4404/4 = 1101$ more lockers. Altogether, there are $1101 + 999 = 2100$ student lockers.

18. **(B)** Without loss of generality, assume A < B < C < D < E. The largest possible sum is $5+6+7+8+9 = 35$, which is not allowed, since A would equal G. The only possible sum of 34 is $4+6+7+8+9$, which again has A=G. No sum is allowed to equal 33, because F would equal G. The sum $4+5+6+8+9 = 32$ has all seven digits different, so 32 is the largest possible sum, and 2 is the corresponding value of G.

19. **(B)** Let $p(x)$ and $q(x)$ represent two cubic polynomials. The x-coordinates of the intersection points are precisely the zeros of the polynomial $p(x) - q(x)$. This polynomial has degree at most two, so it has at most two zeros. Hence, the graphs of the original cubic functions intersect at most twice.

20. **(D)** The first graph is an inverted V-shaped right angle with vertex at (a, b) and the second is a V-shaped right angle with vertex at (c, d). Thus $(a, b), (2, 5), (c, d)$, and $(8, 3)$ are the vertices of a rectangle. Since the diagonals of this rectangle meet at their common midpoint, the x-coordinate of this midpoint must be $(2 + 8)/2 = 5 = (a + c)/2$. Thus $a + c = 10$.

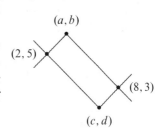

OR

Use the given information to obtain the equations $5 = -|2 - a| + b$, $5 = |2 - c| + d$, $3 = -|8 - a| + b$, and $3 = |8 - c| + d$. Subtract the third from the first to eliminate b and subtract the fourth from the second to eliminate d. The two resulting equations $|8-a|-|2-a| = 2$ and $|2 - c| - |8 - c| = 2$ can be solved for a and c. For example, to solve the former, first consider all $a \leq 2$, for which the equation reduces to $8 - a - (2 - a) = 2$, which has no solutions. Then consider all a in the interval $2 \leq a \leq 8$ for which the equation reduces to $8 - a - (a - 2) = 2$, which yields $a = 4$. Finally, if $a \geq 8$ the equation reduces to $a - 8 - (a - 2) = 2$, which has no solutions. The other equations can be solved similarly to show that $c = 6$ and $a + c = 10$.

21. **(E)** Since I and II cannot hold simultaneously, one of them must be false. Because only one of the four statements is false, statements III and IV are each true, and the correct answer is E.

22. **(C)** The triangle is a right triangle and its hypotenuse is a diameter of the circle, so the region with area C is a semicircle and is congruent to the semicircle formed by the other three regions. The area of the triangle is 6, so $A + B + 6 = C$. To see that the other options are incorrect, note that

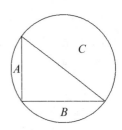

(A) $A + B < A + B + 6 = C$;

(B) $A^2 + B^2 < (A + B)^2 < (A + B + 6)^2 = C^2$;

(D) $4A + 3B < 5A + 5B < 5(A + B + 6) = 5C$; and

(E) $\dfrac{1}{A^2} + \dfrac{1}{B^2} > \dfrac{1}{A^2} > \dfrac{1}{C^2}$.

23. **(D)** The four sums in question are

$$7 + a + b + 1 = K,$$
$$7 + c + 3 = K,$$
$$3 + e + f + 10 = K,$$
$$1 + d + 10 = K.$$

Therefore,

$$4K = 8 + a + b + 10 + c + 13 + e + f + 11 + d$$
$$= 42 + a + b + c + d + e + f$$
$$= 42 + (2 + 4 + 5 + 6 + 8 + 9) = 76$$

It follows that $K = 76/4 = 19$. In fact, $c = 9, d = 8, \{e, f\} = \{2, 4\}$, and $\{a, b\} = \{5, 6\}$ works.

24. **(B)** The area of the triangle is equal to the area of the sector minus the area of the segment. The area of the lune is equal to the area of the semicircle minus the area of the same segment. Thus, we need only find the areas of the sector and the semicircle. They are equal, and the difference between them and the area of the segment is the same. The area of the sector is $\pi \cdot r^2/4$ and the area of the semicircle is $(\pi/2) \cdot \left(r\sqrt{2}/2\right)^2 = \pi \cdot r^2/4$.

25. **(A)** An exterior angle in a regular hexagon is 60° and an exterior angle in a regular pentagon is 72°. Hence the measure of angle CBA is 132°. Since the triangle ABC is isosceles, with $AB = BC$, the measure of angle BAC is $(180° - 132°)/2 = 48°/2 = 24°$.

51st AMC 12 solutions, 2000

1. **(E)** Factor 2001 into primes to get $2001 = 3 \cdot 23 \cdot 29$. The largest possible sum of three distinct factors whose product is 2001 is the one which combines the two largest prime factors, namely $I = 23 \cdot 29 = 667$, $M = 3$, and $O = 1$, so the largest possible sum is $1 + 3 + 667 = 671$.

2. **(A)** $2000(2000^{2000}) = (2000^1)(2000^{2000}) = 2000^{(1+2000)} = 2000^{2001}$. All the other options are greater than 2000^{2001}.

3. **(B)** Since Jenny ate 20% of the jellybeans remaining each day, 80% of the jellybeans are left at the end of each day. If x is the number of jellybeans in the jar originally, then $(0.8)^2 x = 32$. Thus $x = 50$.

4. **(C)** The sequence of units digits is
$$1, 1, 2, 3, 5, 8, 3, 1, 4, 5, 9, 4, 3, 7, 0, 7, 7, 4, 1, 5, 6, \ldots.$$
The digit 6 is the last of the ten digits to appear.

5. **(C)** Since $x < 2$, it follows that $|x - 2| = 2 - x$. If $2 - x = p$, then $x = 2 - p$. Thus $x - p = 2 - 2p$.

6. **(C)** There are five prime numbers between 4 and 18: 5, 7, 11, 13, and 17. Hence the product of any two of these is odd and the sum is even. Because $xy - (x + y) = (x - 1)(y - 1) - 1$ increases as either x or y increases (since both x and y are bigger than 1), the answer must be an odd number that is no smaller than $23 = 5 \cdot 7 - (5 + 7)$ and no larger than $191 = 13 \cdot 17 - (13 + 17)$. The only possibility among the options is 119, and indeed $119 = 11 \cdot 13 - (11 + 13)$.

7. **(E)** If $\log_b 729 = n$, then $b^n = 729 = 3^6$, so n must be an integer factor of 6; that is, $n = 1, 2, 3,$ or 6. Since $729 = 729^1 = 27^2 = 9^3 = 3^6$, the corresponding values of b are $3^6, 3^3, 3^2,$ and 3.

8. **(C)** Calculating the number of squares in the first few figures uncovers a pattern. Figure 0 has $2(0) + 1 = 2(0^2) + 1$ squares, figure 1 has $2(1) + 3 = 2(1^2) + 3$ squares, figure 2 has $2(1 + 3) + 5 = 2(2^2) + 5$ squares, and figure 3 has $2(1 + 3 + 5) + 7 = 2(3^2) + 7$ squares. In general, the number of unit squares in figure n is

$$2(1 + 3 + 5 + \cdots + (2n - 1)) + 2n + 1 = 2(n^2) + 2n + 1.$$

Therefore, the figure 100 has $2(100^2) + 2 \cdot 100 + 1 = 20201.$

OR

Each figure can be considered to be a large square with identical small pieces deleted from each of the four corners. Figure 1 has $3^2 - 4(1)$ unit squares, figure 2 has $5^2 - 4(1 + 2)$ unit squares, and figure 3 has $7^2 - 4 \cdot (1 + 2 + 3)$ unit squares. In general, figure n has

$$(2n + 1)^2 - 4(1 + 2 + \cdots + n) = (2n + 1)^2 - 2n(n + 1) \text{ unit squares.}$$

Thus figure 100 has $201^2 - 200(101) = 20201$ unit squares.

OR

The number of unit squares in figure n is the sum of the first n positive odd integers plus the sum of the first $n + 1$ positive odd integers. Since the sum of the first k positive odd integers is k^2, figure n has $n^2 + (n + 1)^2$ unit squares. So figure 100 has $100^2 + 101^2 = 20201$ unit squares.

9. **(C)** Note that the integer average condition means that the sum of the scores of the first n students is a multiple of n. The scores of the first two students must be both even or both odd, and the sum of the scores of the first three students must be divisible by 3. The remainders when $71, 76, 80, 82,$ and 91 are divided by 3 are $2, 1, 2, 1,$ and 1, respectively. Thus the only sum of three scores divisible by 3 is $76 + 82 + 91 = 249$, so the first two scores entered are 76 and 82 (in some order), and the third score is 91. Since 249 is 1 larger than a multiple of 4, the fourth score must be 3 larger than a multiple of 4, and the only possibility is 71, leaving 80 as the score of the fifth student.

10. **(E)** Reflecting the point $(1, 2, 3)$ in the xy-plane produces $(1, 2, -3)$. A half-turn about the x-axis yields $(1, -2, 3)$. Finally, the translation gives $(1, 3, 3)$.

11. **(E)** Combine the three terms over a common denominator and replace ab in the numerator with $a - b$ to get

$$\frac{a}{b} + \frac{b}{a} - ab = \frac{a^2 + b^2 - (ab)^2}{ab}$$
$$= \frac{a^2 + b^2 - (a - b)^2}{ab}$$
$$= \frac{a^2 + b^2 - (a^2 - 2ab + b^2)}{ab}$$
$$= \frac{2ab}{ab} = 2.$$

OR

Note that $a = a/b - 1$ and $b = 1 - b/a$. It follows that $\frac{a}{b} + \frac{b}{a} - ab = \frac{a}{b} + \frac{b}{a} - \left(\frac{a}{b} - 1\right)\left(1 - \frac{b}{a}\right) = \frac{a}{b} + \frac{b}{a} - \left(\frac{a}{b} + \frac{b}{a} - 2\right) = 2.$

12. **(E)** Note that

$$AMC + AM + MC + CA =$$
$$(A + 1)(M + 1)(C + 1) - (A + M + C) - 1 = pqr - 13,$$

where p, q, and r are positive integers whose sum is 15. A case-by-case analysis shows that pqr is largest when $p = 5$, $q = 5$, and $r = 5$. Thus the answer is $5 \cdot 5 \cdot 5 - 13 = 112$.

13. **(C)** Suppose that the whole family drank x cups of milk and y cups of coffee. Let n denote the number of people in the family. The information given implies that $x/4 + y/6 = (x + y)/n$. This leads to

$$3x(n - 4) = 2y(6 - n).$$

Since x and y are positive, the only positive integer n for which both sides have the same sign is $n = 5$.

OR

If Angela drank c cups of coffee and m cups of milk, then $0 < c < 1$ and $m + c = 1$. The number of people in the family is $6c + 4m = 4 + 2c$, which is an integer if and only if $c = 1/2$. Thus, there are five people in the family.

14. **(E)** If x were less than or equal to 2, then 2 would be both the median and the mode of the list. Thus $x > 2$. Consider the two cases $2 < x < 4$, and $x \geq 4$.

Case 1: If $2 < x < 4$, then 2 is the mode, x is the median, and $(25 + x)/7$ is the mean, which must equal $2 - (x - 2)$, $(x + 2)/2$, or $x + (x - 2)$, depending on the size of the mean relative to 2 and x. These give $x = 3/8$, $x = 36/5$, and $x = 3$, of which $x = 3$ is the only value between 2 and 4.

Case 2: If $x \geq 4$, then 4 is the median, 2 is the mode, and $(25+x)/7$ is the mean, which must be 0, 3, or 6. Thus $x = -25, -4$, or 17, of which 17 is the only one of these values greater than or equal to 4.

Thus the x-values sum to $3 + 17 = 20$.

15. **(B)** Let $x = 9z$. Then $f(3z) = f(9z/3) = f(3z) = (9z)^2 + 9z + 1 = 7$. Simplifying and solving the equation for z yields $81z^2 + 9z - 6 = 0$, so $3(3z + 1)(9z - 2) = 0$. Thus $z = -1/3$ or $z = 2/9$. The sum of these values is $-1/9$. **Note.** The answer can also be obtained by using the sum-of-roots formula on $81z^2 + 9z - 6 = 0$. The sum of the roots is $-9/81 = -1/9$.

16. **(D)** Suppose each square is identified by an ordered pair (m, n), where m is the row and n is the column in which it lies. In the original system, each square (m, n) has the number $17(m-1)+n$ assigned; in the renumbered system, it has the number $13(n-1)+m$ assigned to it. Equating the two expressions yields $4m - 3n = 1$, whose acceptable solutions are $(1, 1), (4, 5), (7, 9), (10, 13)$, and $(13, 17)$. These squares are numbered 1, 56, 111, 166, and 221, respectively, and the sum is 555.

17. **(D)** The fact that $OA = 1$ implies that $BA = \tan \theta$ and $BO = \sec \theta$. Since $\overline{BC}$ bisects $\angle ABO$, it follows that $OB/BA = OC/CA$, which implies

$$\frac{OB}{OB + BA} = \frac{OC}{OC + CA} = OC.$$

Substituting yields

$$OC = \frac{\sec \theta}{\sec \theta + \tan \theta} = \frac{1}{1 + \sin \theta}.$$

OR

Let $\alpha = \angle CBO = \angle ABC$. Using the *Law of Sines* on triangle BCO

yields $\sin\theta/BC = \sin\alpha/OC$, so $OC = BC\sin\alpha/\sin\theta$. In right triangle ABC, $\sin\alpha = (1-OC)/BC$. Hence $OC = (1-OC)/\sin\theta$. Solving this for OC yields $OC = 1/(1+\sin\theta)$.

18. **(A)** Note that, if a Tuesday is d days after a Tuesday, then d is a multiple of 7. Next, we need to consider whether any of the years $N-1, N, N+1$ is a leap year. If N is not a leap year, the 200th day of year $N+1$ is $365 - 300 + 200 = 265$ days after a Tuesday, and thus is a Monday, since 265 is 6 larger than a multiple of 7. Thus, year N is a leap year and the 200th day of year $N+1$ is another Tuesday (as given), being 266 days after a Tuesday. It follows that year $N-1$ is not a leap year. Therefore, the 100th day of year $N-1$ precedes the given Tuesday in year N by $365 - 100 + 300 = 565$ days, and therefore is a Thursday, since $565 = 7 \cdot 80 + 5$ is 5 larger than a multiple of 7.

OR

This solution uses the notation of modular arithmetic. Note that if a Tuesday is d days after a Tuesday, then $d \equiv 0 \pmod{7}$ (see note below). Next, we need to consider whether any of the years $N-1, N, N+1$ is a leap year. If N is not a leap year, the 200th day of year $N+1$ is $365 - 300 + 200 = 265$ days after a Tuesday, and thus is a Monday, since $265 \equiv 6 \pmod{7}$. If N is a leap year, the 200th day of year $N+1$ is 266 days after a Tuesday, and thus is another Tuesday, as given. It follows that N is a leap year, and that $N-1$ is not a leap year. The 100th day of year $N-1$ precedes a Tuesday in year N by $365 - 100 + 300 = 565$ days, and thus is a Thursday, since $565 \equiv 5 \pmod{7}$. **Note.** To say u is congruent to $i \pmod{n}$ means that $u - i$ is divisible by n. This relationship is written $u \equiv i \pmod{n}$.

19. **(C)** Since the edges of the triangle are known, we can use Heron's Formula to find the area. The area of triangle ABC is $\sqrt{(21)(8)(7)(6)}$, which is 84, so the altitude from vertex A is $2(84)/14 = 12$. The midpoint D divides $\overline{BC}$ into two segments of length 7, and the bisector of angle BAC divides $\overline{BC}$ into segments of length $14(13/28) = 6.5$ and $14(15/28) = 7.5$ (since the angle bisector divides the opposite side into lengths proportional to the remaining two

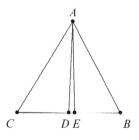

sides). Thus the triangle ADE has base $DE = 7 - 6.5 = 0.5$ and altitude 12, so its area is 3.

20. **(B)** Note that $(x+1/y)+(y+1/z)+(z+1/x) = 4+1+7/3 = 22/3$ and that

$$28/3 = 4 \cdot 1 \cdot 7/3 = (x + 1/y)(y + 1/z)(z + 1/x)$$
$$= xyz + x + y + z + 1/x + 1/y + 1/z + 1/(xyz)$$
$$= xyz + 22/3 + 1/(xyz).$$

It follows that $xyz + 1/(xyz) = 2$ and $(xyz - 1)^2 = 0$. Hence $xyz = 1$.

OR

By substitution,

$$4 = x + \frac{1}{y} = x + \frac{1}{1 - 1/z} = x + \frac{1}{1 - 3x/(7x - 3)} = x + \frac{7x - 3}{4x - 3}.$$

Thus $4(4x-3) = x(4x-3)+7x-3$, which simplifies to $(2x-3)^2 = 0$. Accordingly, $x = 3/2, z = 7/3-2/3 = 5/3$, and $y = 1-3/5 = 2/5$, so $xyz = (3/2)(2/5)(5/3) = 1$.

21. **(D)** Without loss of generality, let the side of the square have length 1 unit and let the area of triangle ADF be m. Let $AD = r$ and $EC = s$. Because triangles ADF and FEC are similar, $s/1 = 1/r$. Since $\frac{1}{2}r = m$, the area of triangle FEC is $s/2 = 1/2r = 1/4m$.

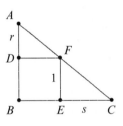

OR

Let $B = (0,0)$, $E = (1,0)$, $F = (1,1)$, and $D = (0,1)$ be the vertices of the square. Let $C = (1 + 2m, 0)$, and notice that the area of $BEFD$ is 1 and the area of triangle FEC is m. The slope of the line through C and F is $-1/2m$; thus, it intersects the y-axis at $A = (0, 1 + 1/2m)$. The area of triangle ADF is therefore $1/4m$.

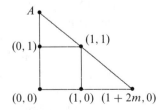

22. **(C)** First note that the quartic polynomial can have no more real zeros than the two shown. (If it did, the quartic $P(x) - 5$ would have more than four zeros.) The sum of the coefficients of P is $P(1)$, which is greater than 3. The product of all the zeros of P is the constant term of the polynomial, which is the y-intercept, which is greater than 5. The sum of the real zeros of P (the sum of the x-intercepts) is greater than 4.5, and $P(-1)$ is greater than 4. However, since the product of the real zeros of P is greater than 4.5 and the product of all the zeros is less than 6, it follows that the product of the non-real zeros of P is less than 2, making it the smallest of the numbers.

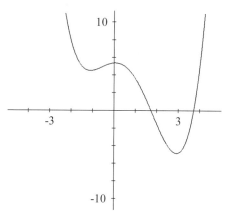

23. **(B)** In order for the sum of the logarithms of six numbers to be an integer k, the product of the numbers must be 10^k. The only prime factors of 10 are 2 and 5, so the six integers must be chosen from the list 1, 2, 4, 5, 8, 10, 16, 20, 25, 32, 40. For each of these, subtract the number of times that 5 occurs as a factor from the number of times 2 occurs as a factor. This yields the list 0, 1, 2, −1, 3, 0, 4, 1, −2, 5, 2. Because 10^k has just as many factors of 2 as it has of 5, the six chosen integers must correspond to six integers in the latter list that sum to 0. Two of the numbers must be −1 and −2, because there are only two zeros in the list, and no number greater than 2 can appear in the sum, which must therefore be $(-2) + (-1) + 0 + 0 + 1 + 2 = 0$. It follows that Professor Gamble chose 25, 5, 1, 10, one number from $\{2, 20\}$, and one number from $\{4, 40\}$. There are four possible tickets Professor Gamble could have bought and only one is a winner, so the probability that Professor Gamble wins the lottery is $1/4$.

OR

As before, the six integers must be chosen from the set $S = \{1, 2, 4, 5, 8, 10, 16, 20, 25, 32, 40\}$. The product of the smallest six numbers in S is $3{,}200 > 10^3$, so the product of the numbers on the ticket must be 10^k for some $k \geq 4$. On the other hand, there are only six factors of 5 available among the numbers in S, so the product p can only be 10^4, 10^5, or 10^6.

Case 1, $p = 10^6$. There is only one way to produce 10^6, since all six factors of 5 must be used and their product is already 10^6, leaving 1 as the other number: $1, 5, 10, 20, 25, 40$.

Case 2, $p = 10^5$. To produce a product of 10^5 we must use six numbers that include five factors of 5 and five factors of 2 among them. We cannot use both 20 and 40, because these numbers combine to give five factors of 2 among them and the other four numbers would have to be odd (whereas there are only three odd numbers in S). If we omit 40, we must include the other multiples of 5 (5, 10, 20, 25) plus two numbers whose product is 4 (necessarily 1 and 4). If we omit 20, we must include 5, 10, 25, and 40, plus two numbers with a product of 2 (necessarily 1 and 2).

Case 3, $p = 10^4$. To produce a product of 10^4 we must use six numbers that include four factors of 5 and four factors of 2 among them. So that there are only four factors of 2, we must include 1, 5, 25, 2, and 10. These include two factors of 2 and four factors of 5, so the sixth number must contain two factors of 2 and no 5's, so must be 4.

Thus there are four lottery tickets whose numbers have base-ten logarithms with an integer sum. They are $\{1, 5, 10, 20, 25, 40\}$, $\{1, 2, 5, 10, 25, 40\}$, $\{1, 2, 4, 5, 10, 25\}$, and $\{1, 4, 5, 10, 20, 25\}$. Professor Gamble has a $1/4$ probability of being a winner.

24. **(D)** Construct the circle with center A and radius AB. Let F be the point of tangency of the two circles. Draw $\overline{AF}$, and let E be the point of intersection of $\overline{AF}$ and the given circle. By the *Power of a Point Theorem*, $AD^2 = AF \cdot AE$ (see Note below). Let r be the radius of the smaller circle. Since $\overline{AF}$ and $\overline{AB}$ are radii of the larger circle, $AF = AB$ and $AE = AF - EF = AB - 2r$. Because $AD = AB/2$, substitution into the first equation yields

$$(AB/2)^2 = AB \cdot (AB - 2r),$$

or, equivalently, $\dfrac{r}{AB} = \dfrac{3}{8}$. Points A, B, and C are equidistant from each other, so $\overparen{BC} = 60°$ and thus the circumference of the larger circle is $6 \cdot$ (length of $\overparen{BC}) = 6 \cdot 12$. Let c be the circumference of the smaller circle. Since the circumferences of the two circles are in the same ratio as their radii, $\frac{c}{72} = \frac{r}{AB} = \frac{3}{8}$. Therefore

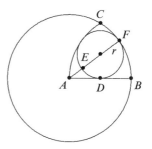

$$c = \frac{3}{8}(72) = 27.$$

Note. From any exterior point P, a secant PAB and a tangent PT are drawn. Consider triangles PAT and PTB. They have a common angle P. Since angles ATP and PBT intercept the same arc $\overparen{AT}$, they are congruent. Therefore triangles PAT and PTB are similar, and it follows that $PA/PT = PT/PB$ and $PA \cdot PB = PT^2$. The number PT^2 is called *the power of the point* P with respect to the circle. Intersecting secants, tangents, and chords, paired in any manner create various cases of this theorem, which is sometimes called *Crossed Chords*.

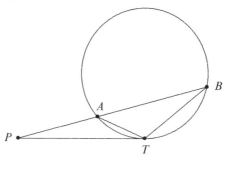

OR

The construction given in the problem is the classic way to construct an equilateral triangle, $\triangle ABC$, with side length AB. The arc length BC is one-sixth the circumference of the circle with radius AB, so

$$12 = \frac{1}{6}(2\pi \cdot AB) \quad \text{and} \quad AB = \frac{36}{\pi}.$$

Let O be the center of the circle, r be the radius, and D be the midpoint of AB. The symmetry of the region implies that $\overline{OD}$ is a perpendicular bisector of AB. Construct $\overline{AE}$, the line segment passing through O and intersecting the arc $\overline{BC}$ at E. Then $AE = AB$ and $r = OE = OD$, so in the right $\triangle ADO$ we have

$$\frac{36}{\pi} = AE = OE + AO = OE + \sqrt{AD^2 + DO^2}$$

$$= r + \sqrt{\left(\frac{18}{\pi}\right)^2 + r^2}.$$

Hence

$$0 = \left(\frac{36}{\pi} - r\right)^2 - \left(\left(\frac{18}{\pi}\right)^2 + r^2\right)$$

$$= 3\left(\frac{18}{\pi}\right)^2 - \frac{72}{\pi}r.$$

and

$$r = \frac{\pi}{72} \cdot 3\left(\frac{18}{\pi}\right)^2 = \frac{27}{2\pi}.$$

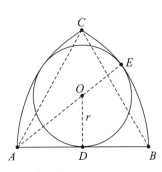

The circumference of the circle is $2\pi r = 2\pi\left(\frac{27}{2\pi}\right) = 27$.

25. **(E)** The octahedron has eight congruent equilateral triangular faces that form four pairs of parallel faces. Choose one color for the bottom face. There are seven choices for the color of the top face. Three of the remaining faces have an edge in common with the bottom face. There are $\binom{6}{3} = 20$ ways of choosing the colors for these faces and two ways to arrange these on the three faces (clockwise and counterclockwise). Finally, there are $3! = 6$ ways to fix the last three colors. Thus the total number of distinguishable octahedrons that can be constructed is $7 \cdot 20 \cdot 2 \cdot 6 = 1680$.

OR

Place a cube inside the octahedron so that each of its vertices touches a face of the octahedron. Then assigning colors to the faces of the octahedron is equivalent to assigning colors to the vertices of the cube. Pick one vertex and assign it a color. Then the remaining colors can be assigned in 7! ways.

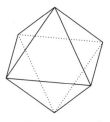

Since three vertices are joined by edges to the first vertex, they are interchangeable by a rotation of the cube, hence the answer is $7!/3 = 1680$.

1st AMC 10 solutions, 2000

1. **(E)** Factor 2001 into primes to get $2001 = 3 \cdot 23 \cdot 29$. The largest possible sum of three distinct factors whose product is 2001 is the one which combines the two largest prime factors, namely $I = 23 \cdot 29 = 667, M = 3$, and $O = 1$, so the largest possible sum is $1 + 3 + 667 = 671$.

2. **(A)** $2000(2000^{2000}) = (2000^1)(2000^{2000}) = 2000^{(1+2000)} = 2000^{2001}$. All the other options are greater than 2000^{2001}.

3. **(B)** Since Jenny ate 20% of the jellybeans remaining each day, 80% of the jellybeans are left at the end of each day. If x is the number of jellybeans in the jar originally, then $(0.8)^2 x = 32$. Thus $x = 50$.

4. **(D)** Since Chandra paid an extra $5.06 in January, her December connect time must have cost her $5.06. Therefore, her monthly fee is $12.48 - \$5.06 = \$7.42$.

5. **(B)** By the *Triangle Midsegment Theorem*, $MN = AB/2$. Since the base AB and the altitude to AB of $\triangle ABP$ do not change, the area does not change. The altitude of the trapezoid is half that of the triangle, and the bases do not change as P changes, so the area of the trapezoid does not change. Only the perimeter changes (reaching a minimum when $\triangle ABP$ is isosceles).

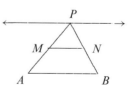

145

6. **(C)** The sequence of units digits is
$$1, 1, 2, 3, 5, 8, 3, 1, 4, 5, 9, 4, 3, 7, 0, 7, 7, 4, 1, 5, 6, \ldots.$$
The digit 6 is the last of the 10 digits to appear.

7. **(B)** Both triangles APD and CBD are $30°$–$60°$–$90°$ triangles. Thus $DP = 2\sqrt{3}/3$ and $DB = 2$. Since $\angle BDP = \angle PBD$, it follows that $PB = PD = 2\sqrt{3}/3$. Hence the perimeter of $\triangle BDP$ is
$$2\sqrt{3}/3 + 2\sqrt{3}/3 + 2 = 2 + 4\sqrt{3}/3.$$

8. **(D)** Let f and s represent the numbers of freshmen and sophomores at the school, respectively. According to the given condition, $(2/5)f = (4/5)s$. Thus, $f = 2s$. That is, there are twice as many freshmen as sophomores.

9. **(C)** Since $x < 2$, it follows that $|x - 2| = 2 - x$. If $2 - x = p$, then $x = 2 - p$. Thus $x - p = 2 - 2p$.

10. **(D)** By the *Triangle Inequality*, each of x and y can be any number strictly between 2 and 10, so $0 \le |x - y| < 8$. Therefore, the smallest positive number that is not a possible value of $|x - y|$ is $10 - 2 = 8$.

11. **(C)** There are five prime numbers satisfying the conditions: 5, 7, 11, 13, and 17. Hence the product of any two of these is odd and the sum is even. Because $xy - (x + y) = (x - 1)(y - 1) - 1$ increases as either x or y increases, the answer must be a number that is no smaller than $23 = 5 \cdot 7 - (5 + 7)$ and no larger than $191 = 13 \cdot 17 - (13 + 17)$. The only possibility among the options is 119 which is $11 \cdot 13 - (11 + 13)$.

12. **(C)** Calculating the number of squares in the first few figures uncovers a pattern.

fig 0: $2(0) + 1$ $\qquad = 2(0^2) + 1$
fig 1: $2(1) + 3$ $\qquad = 2(1^2) + 3$
fig 2: $2(1 + 3) + 5$ $\qquad = 2(2^2) + 5$
fig 3: $2(1 + 3 + 5) + 7$ $\qquad = 2(3^2) + 7$
fig 4: $2(1 + 3 + 5 + 7) + 9$ $\qquad = 2(4^2) + 9$
fig n: $2(1 + 3 + 5 + \cdots + (2n - 1)) + 2n + 1 = 2n^2 + 2n + 1$

Therefore the 100th figure has $2(100^2) + 2 \cdot 100 + 1 = 20201$.

OR

Each figure can be considered as a large square with identical small pieces deleted from each of the four corners. Figure 1 has $3^2 - 4(1)$ unit squares, figure 2 has $5^2 - 4(1+2)$ unit squares, and figure 3 has $7^2 - 4 \cdot (1+2+3)$ unit squares. In general, the figure n has
$$(2n+1)^2 - 4(1+2+\cdots+n) = (2n+1)^2 - 2n(n+1).$$
Thus figure 100 will consist of $201^2 - 200(101) = 20201$ unit squares.

OR

The number of unit squares in the nth figure is the sum of the first n positive odd integers plus the sum of the first $n+1$ positive odd integers. Since the sum of the first k positive odd integers is k^2, the number of unit squares in the nth figure is $n^2 + (n+1)^2$. So the number of unit squares in the 100th figure is $100^2 + 101^2 = 20201$.

13. **(B)** To avoid having two yellow pegs in the same row or column, there must be exactly one yellow peg in each row and in each column. Hence the peg in the first row must be yellow, the second peg of the second row must be yellow, the third peg of the third row must be yellow, etc. To avoid having two red pegs in some row, there must be a red peg in each of rows 2, 3, 4, and 5. The red pegs must be in the first position of the second row, the second position of the third row, etc. Continuation yields exactly one ordering that meets the requirements, as shown.

14. **(C)** Note that the integer average condition means that the sum of the scores of the first n students is congruent to 0 (mod n) (see Note below). The scores of the first two students must be both even or both odd, and the sum of the scores of the first three students must be divisible by 3. The remainders when 71, 76, 80, 82, and 91 are divided by 3 are 2, 1, 2, 1, and 1, respectively. Thus the only sum of three scores divisible by 3 is $76 + 82 + 91 = 249$, which is congruent to 1 (mod 4). So the score of the fourth student must be congruent to 3 (mod 4), and the only possibility is 71, leaving 80 as the score of the fifth student. **Note.** To say u is congruent to i (mod n) means that $u - i$ is divisible by n.

15. **(E)** Find the common denominator and replace the ab in the numerator with $a-b$ to get

$$\frac{a}{b}+\frac{b}{a}-ab = \frac{a^2+b^2-(ab)^2}{ab}$$

$$= \frac{a^2+b^2-(a-b)^2}{ab}$$

$$= \frac{a^2+b^2-(a^2-2ab+b^2)}{ab}$$

$$= \frac{2ab}{ab} = 2.$$

OR

Note that $a = a/b - 1$ and $b = 1 - b/a$. It follows that $\frac{a}{b}+\frac{b}{a}-ab =$
$\frac{a}{b}+\frac{b}{a}-\left(\frac{a}{b}-1\right)\left(1-\frac{b}{a}\right) = \frac{a}{b}+\frac{b}{a}-\left(\frac{a}{b}+\frac{b}{a}-2\right) = 2.$

16. **(B)** Extend $\overline{DC}$ to F. Triangles FAE and DBE are similar with ratio $5:4$. Thus $AE = 5 \cdot AB/9$, $AB = \sqrt{3^2+6^2} = \sqrt{45} = 3\sqrt{5}$, and $AE = 5(3\sqrt{5})/9 = 5\sqrt{5}/3$.

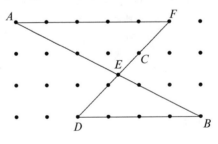

OR

Coordinatize the points so that $A = (0,3)$, $B = (6,0)$, $C = (4,2)$, and $D = (2,0)$. Then the line through A and B is given by $x + 2y = 6$, and the line through C and D is given by $x - y = 2$. Solve these simultaneously to get $E = (10/3, 4/3)$. Hence $AE = \sqrt{(10/3-0)^2+(4/3-3)^2} = \sqrt{125/9} = 5\sqrt{5}/3$.

17. **(D)** Neither of the exchanges *quarter* → *five nickels* nor *nickel* → *five pennies* changes the value of Boris's coins. The exchange *penny* → *five quarters* increases the value of Boris's coins by $1.24. Hence, Boris must have $.01 + $1.24n after n uses of the last exchange. Only

option D is of this form: $745 = 1 + 124 \cdot 6$. In cents, option A is 115 more than a multiple of 124, B is 17 more than a multiple of 124, C is 10 more than a multiple of 124, and E is 39 more than a multiple of 124.

18. **(C)** At any point on Charlyn's walk, she can see all the points inside a circle of radius 1 km. The portion of the viewable region inside the square consists of the interior of the square except for a smaller square with side length 3 km. This portion of the viewable region has area $(25 - 9)$ km^2. The portion of the viewable region outside the square consists of four rectangles, each 5 km by 1 km, and four quarter-circles, each with a radius of 1 km. This portion of the viewable region has area $4(5 + \frac{\pi}{4}) = (20 + \pi)$ km^2. The area of the entire viewable region is $36 + \pi \approx 39$ km^2.

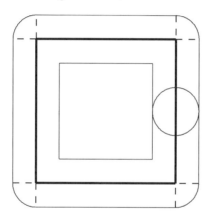

19. **(C)** Notice that $AMC + AM + MC + CA = (A + 1)(M + 1)(C + 1) - (A + M + C) - 1 = pqr - 11$, where p, q, and r are positive integers whose sum is 13. A simple case analysis shows that pqr is largest when two of the numbers p, q, r are 4 and the third is 5. Thus the answer is $4 \cdot 4 \cdot 5 - 11 = 69$.

20. **(B)** From the conditions we can conclude that some creepy crawlers are ferocious (since some are alligators). Hence, there are some ferocious creatures that are creepy crawlers, and thus II must be true. The diagram below shows that the only conclusion that can be drawn is existence of an animal in the region with the dot. Thus, neither I nor III follows from the given conditions.

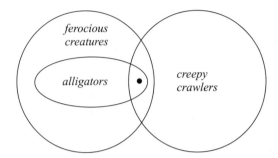

21. **(C)** Suppose that the whole family drinks x cups of milk and y cups of coffee. Let n denote the number of people in the family. The information given implies that $x/4 + y/6 = (x+y)/n$. This leads to
$$3x(n-4) = 2y(6-n).$$
Since x and y are positive, the only positive integer n for which both sides have the same sign is $n = 5$.

OR

If Walter drinks c cups of coffee and m cups of milk, then $0 < c < 1$ and $m + c = 1$. The number of people in the family is $6c + 4m = 4 + 2c$, which is an integer if and only if $c = \frac{1}{2}$. Thus there are five people in the family.

22. **(E)** If x were less than or equal to 2, then 2 would be both the median and the mode of the list. Thus $x > 2$. Consider the two cases $2 < x < 4$, and $x \geq 4$.

Case 1: If $2 < x < 4$, then 2 is the mode, x is the median, and $(25+x)/7$ is the mean, which must equal $2 - (x-2)$, $(x+2)/2$, or $x + (x-2)$, depending on the size of the mean relative to 2 and x. These give $x = 3/8$, $x = 36/5$, and $x = 3$, of which $x = 3$ is the only value between 2 and 4.

Case 2: If $x \geq 4$, then 4 is the median, 2 is the mode, and $(25+x)/7$ is the mean, which must be $0, 3$, or 6. Thus $x = -25, -4$, or 17, of which 17 is the only one of these values greater than or equal to 4.

Thus the x-values sum to $3 + 17 = 20$.

23. **(B)** Let $x = 9z$. Then $f(9z/3) = f(3z) = 81z^2 + 9z + 1 = 7$. Simplifying and solving the equation for z yields $81z^2 + 9z - 6 = 0$,

so $3(3z+1)(9z-2) = 0$. Thus $z = -1/3$ or $z = 2/9$. The sum of these values is $-1/9$. **Note.** The answer could also be obtained by using the sum-of-roots formula on $81z^2 + 9z - 6 = 0$. The sum of the roots is $-9/81 = -1/9$.

24. **(A)** Note that if a Tuesday is d days after a Tuesday, then $d \equiv 0 \pmod{7}$ (see note below). Next, we need to consider whether any of the years $N-1, N, N+1$ is a leap year. If N is not a leap year, the 200th day of year $N+1$ is $365 - 300 + 200 = 265$ days after a Tuesday, and thus is a Monday, since $265 \equiv 6 \pmod 7$. If N is a leap year, the 200th day of year $N+1$ is 266 days after a Tuesday, and thus is another Tuesday, as given. It follows that N is a leap year, and that $N-1$ is not a leap year. The 100th day of year $N-1$ precedes a Tuesday in year N by $365 - 100 + 300 = 565$ days, and thus is a Thursday, since $565 \equiv 5 \pmod 7$. **Note.** To say u is congruent to i (mod n) means that $u - i$ is divisible by n. This relationship is written $u \equiv i \pmod n$.

25. **(D)** Without loss of generality, let the side of the square have length 1 unit and let the area of triangle ADF be m. Let $AD = r$ and $EC = s$. Because triangles ADF and FEC are similar, $s/1 = 1/r$. Since $\frac{1}{2}r = m$, the area of triangle FEC is $s/2 = 1/2r = 1/4m$.

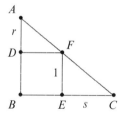

Additional Problems

1. **Dinner Bill Splitting.** Years ago, my neighbors agreed to celebrate our wedding anniversary with my wife and me. The four of us went to a lovely restaurant, enjoyed a fine dinner, and asked for the bill. When it came, we asked that it be split in half. Realizing the waiter's discomfort, we all set to work on the problem. The bill was for an odd amount, so it could not be split perfectly. However, we realized that, except for the penny problem, we could take half the bill by simply reversing the dollars and the cents. In other words, if we double t dollars and s cents, the result differs by 1 cent from s dollars and t cents. We told the waiter about this. He was astounded: "I never knew you could do it that way." Later, over another dinner with mathematical friends, the question of uniqueness came up, and pretty soon we realized that this number is the only one with this surprising splitting property. What was the amount of the original bill?

2. **The 7-11 Problem.** A man goes into a convenience store, picks out four items, and goes to check out. The cashier tells him that her cash register is broken, and she will use her calculator. She proceeds to process the four amounts, and says, "that will be $7.11." "Wait a minute", he protests, "you multiplied the prices together." She promptly repeats the calculation, this time adding the four amounts, and exclaims, "There, you owe $7.11, just as I said." (There is no tax on food in this state.) There are two questions. First, what is the name of the convenience store, and what are the four prices? Challenge: try this problem with only three items. You'll have to change the $7.11, of course. Then try the problem for just two items. There are lots of

solutions. Find them all. Then try the 7-11 problem with three items and a total bill of $8.25. Find some other total cost that could be used to solve the three item 7-11 problem.

3. **Longest Path.** Each rectangle in the diagram is 2×1. What is the length of the longest path from A to B along the edges that does not retrace any of its edges? Prove that your answer is a maximum.

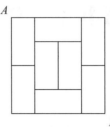

4. **The Wizards.** This captivating problem is due to John Conway, Princeton University. Two wizards get on a bus, and one says to the other 'I have a positive number of children each of whom is a positive integer number of years old. The sum of their ages is the number of this bus and the product of their ages is my age.' To this the second wizard replies 'perhaps if you told me your age and the number of children, I could work out their individual ages.' The first wizard then replies 'No, you could not.' Now the second wizard says 'Now I know your age.' What is the number of the bus? Note: Wizards reason perfectly, can have any number of children, and can be any positive integer years old. Also, consider the same problem but with the additional assumption that the children are all different ages.

5. **Pebbling.** (M Kontsevich, 1981 Tournament of Towns). The first quadrant is decomposed into squares for the following game. Some of these squares are occupied by counters. A position with counters may be transformed to another position according to the following rule: If the neighboring squares to the right and above a counter are both free, it is possible to remove the counter and replace it with counters at both these free squares. The goal is to have all the shaded squares free of counters. Is it possible to reach this goal if the initial position has just one counter in the lower left-hand corner?

Additional Problems 155

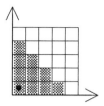

6. **Thirty Digits.** Prove that in every 30-digit number of the form $2^i 5^j 7^k$, some digit appears at least 4 times in its decimal representation.

7. **The Whispered Number Problems.** This pair of problems came to me from Wen-Hsien Sun. Version A was on the Fifth Po Leung Kuk Primary Math World Contest of Taiwan. Version B. is probably the one that was intended.

   Version A. A teacher whispers a positive integer p to student P, a positive integer q to student Q, and a positive integer r to student R. The students don't know one another's numbers but they know the sum of the three numbers is 14. The students make the following statements:

   (a) P says 'I know that Q and R have different numbers.'

   (b) Q says 'I already knew that all three of our numbers are different.'

   (c) R says 'Now I know all three of our numbers.'

   What is the product of the three numbers?

   Version B. A teacher whispers a positive integer p to student P, q to student Q, and r to student R. The students don't know one another's numbers but they know the sum of the three numbers is 14. The students make the following statements:

   (a) P says 'I know that Q and R have different numbers.'

   (b) Q says '<u>Now I know</u> that all three of our numbers are different.'

   (c) R says 'Now I know all three of our numbers.'

   Assume all three students reason perfectly. What is the product of the three numbers?

8. **The Staircase Problem.** Rick Armstrong, St Louis Community College, posed and solved the following problem in Mathematics and Computer Education, (http://www.macejournal.org/) (AN-1, Fall, 2001, and (solution) page 108 of the Spring 2003 issue). An

n-staircase is a grid of $1 + 2 + \cdots + n = \binom{n+1}{2}$ squares arranged so that column 1 has 1 square, column 2 has 2 squares, ..., and column n has n squares. How many *square* regions are bounded by the gridlines of an n-staircase. The question posed here is "How many *rectangular* regions are bounded by the gridlines of an n staircase?" The figure shows a 5-staircase.

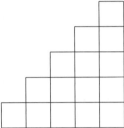

9. **Piles of Stones.** This beauty comes from the Spring 2001 Tournament of Towns contest. Three piles of stones contain 5, 49, and 51 stones. Two operations are allowed: (a) any two piles can be joined together into one pile and (b) any pile with an even number of stones can be divided into two piles of equal size. Is it possible to use the two operations to achieve 105 piles each with one stone?

10. **Flipping Pennies.** A table has some coins on it. Each one shows either heads or tails. You are told the total number of heads showing. Without looking at the status of any coin, is it possible to divide the coins into two groups, perhaps turning some over, so that each of the two groups has the same number of heads showing?

11. **Back to Front.** Given a stack of eleven cards numbered $11, 10, 9, \ldots, 1$, we wish to reverse their order to give $1, 2, 3, \ldots, 11$. To do this we are allowed at any stage to make a *move* of the following type: Remove any section of adjacent cards from the pack and insert them elsewhere in the pack. For example, one initial move is to reposition 9, 8, 7 to give the ordering $11, 10, 6, 5, 9, 8, 7, 4, 3, 2, 1$. What is the minimum number of moves required to reverse the 11 cards?

12. **Face Painting.** Suppose some faces of a large wooden cube are painted red and the rest are painted black. The cube is then cut into unit cubes. The number of unit cubes with some red paint is found to be exactly 200 larger than the number of cubes with some black paint. How many cubes have no paint at all?

Additional Problems

13. **High Slopes.** The numbers $1, 2, 3, \ldots, 100$ are arranged in a 10×10 grid so that consecutive numbers occupy adjacent squares. What is the greatest possible sum of the numbers along a diagonal of the grid?

14. **An Odd End.** Let n denote a positive integer. Let $P(n)$ denote the product of the ten numbers consisting of n and the next nine consecutive numbers. For example, $P(11) = 11 \cdot 12 \cdot 13 \cdots 20 = 670442572800$. The rightmost digit of $P(n)$ is always zero. Notice that the last digit before the zeros of $P(11)$ is even, and more often than not, this is the case. For some integers n, however, the last non-zero digit of $P(n)$ is odd. What is the least such n?

15. **Prime Leaps.** The history/math teacher asked the class to name some years that they knew from history lessons. Johnny named 1066, the Battle of Hastings, and 1939, the outbreak of World War II. The teacher then asked him to calculate the number of years between the two events, and Johnny correctly answered 873. The teacher then asked Johnny if that difference is a prime number and Johnny correctly answered that it is not prime since it is divisible by 3. The teacher than asked the class to find the longest list of years from $0, 1, 2, 3, 4, \ldots, 1996$ so that any two numbers in the list have a difference that is not prime. What numbers are in the longest such list?

16. **Multiple Quotients.** Parentheses can be inserted into the expression $1 \div 2 \div 3 \div 4$ in various ways. For example, $(1 \div 2) \div (3 \div 4) = 2/3$, whereas $1 \div ((2 \div 3) \div 4) = 6$. Similarly, brackets can be inserted into $1 \div 2 \div 3 \div 4 \div 5 \div 6 \div 7 \div 8 \div 9 \div 10 \div 11$ to produce a large collection of whole numbers. What is the ratio of the largest of these whole numbers to the smallest of these whole numbers?

17. **Unsquare Party.** Ashley noticed that the set of ages of her relatives, all of whom were whole numbers in the range 1 up to 100 inclusive, has the unusual property that no two of them multiplied together is a perfect square. What is the largest number of relatives Ashley could have?

18. **Allison's Coin Machine.** Allison has an incredible coin machine. When she puts in a nickel, it gives back five pennies, and when she puts in a penny, it gives back five nickels. If she starts with just

one penny, is it possible that she will ever have the same number of pennies and nickels?

19. **Chameleons.** On the island of Camelot live 45 chameleons, 13 of which are grey, 15 of which are brown, and 17 of which are crimson. If two chameleons of different colors meet, they both simultaneously change to the third color. Is it possible that they will all eventually be the same color?

20. **Repeated Arithmetic.** (University of Maryland High School Math Competition, 1997.) There are 2003 nonzero real numbers written on a blackboard. An operation consists of choosing any two of these, a and b, erasing them, and writing $a + b/2$ and $b - a/2$ in their places. Prove that no sequence of operations can return the set of numbers to the original set.

21. **Double or Add Seven.** (This nice problem came to me from Steve Blasberg.) A collection S of numbers is defined as follows:
    (a) 2 is in S.
    (b) if n is in S, then $2n$ is also in S.
    (c) if n is in S, then $n + 7$ is also in S.
    (d) no other numbers belong to S.
    What is the smallest number larger than 2004 that is NOT in S?

22. **Double or Subtract Twelve.** (I received this problem from Rick Armstrong.) Define a function f as follows:

$$f(n) = \begin{cases} n - 12 & \text{if } n > 25, \\ 2n & \text{if } n \leq 25. \end{cases}$$

How many of the first 1000 positive integers n have the property that $f^k(n) = \underbrace{f \circ f \circ \cdots \circ f}_{k}(n) = 16$ for some positive integer k?

23. **Counting Transitive Relations.** A relation R on a set A is a set of ordered pairs of members of A. That is, R is a relation on A if $R \subset A \times A$. A relation R is called *transitive* if for all $x, y, z \in A, (x, y) \in R$ and $(y, z) \in R$, then $(x, z) \in R$. Let $A = \{1, 2, 3\}$. One relation on A is the empty relation, which is transitive because the implication above is satisfied vacuously. How many of the $2^9 = 512$ relations on A are transitive?

Solutions to Additional Problems

1. Dinner Bill Splitting Problem

Solution. First, we'll state the problem in a more precise way. Twice t dollars and s cents differs by just one cent from s dollars and t cents. Find s and t. In terms of cents,
$$|100s + t - 2(100t + s)| = 1.$$
This is equivalent to $|98s - 199t| = 1$ which can be interpreted as the two equations $98s - 199t = 1$ or $98s - 199t = -1$.

Before continuing with the problem at hand, consider another problem whose solution will propel us toward solving the one at hand. The Decanting Problem is a liquid measuring problem that begins with two unmarked decanters with capacities a and b. Usually a and b are integers. The problem is to determine the smallest amount of liquid that can be measured and how such amount can be obtained, by a process of filling, pouring, and dumping. Specifically, there are three actions we can take:
1. fill an empty decanter,
2. dump out a full decanter, and
3. pour from one decanter to the other until either the receiving decanter is full or the poured decanter is empty.

Let's look at an easy one first. Let $a = 3$ and $b = 5$. We can fill the 3 unit decanter twice, and dump the 5 unit decanter once to get 1 unit of liquid. Algebraically, $2 \cdot 3 - 1 \cdot 5 = 1$. Next, suppose the decanters have capacities 5 units and 7 units. A little experimentation leads to the conclusion that 1 unit of water can be obtained by filling the 5 unit decanter 3 times, pouring repeatedly from the 5 unit to the 7 unit decanter and dumping out the 7 unit decanter twice. A finite state diagram is helpful to follow the procedure:

$$(0,0) \Longrightarrow (5,0) \Longrightarrow (0,5) \Longrightarrow (5,5) \Longrightarrow (3,7) \Longrightarrow (3,0)$$
$$\Longrightarrow (0,3) \Longrightarrow (5,3) \Longrightarrow (1,7) \Longrightarrow (1,0),$$

where the notation (x, y) means the 5-unit container has x units of liquid and the 7-unit container has y units. Notice that the procedure includes 3 fills and 2 dumps, with fills and dumps alternating and separated by 4 pours. An arithmetic equation representing this is

$$3 \cdot 5 - 2 \cdot 7 = 1.$$

Notice that not only does the arithmetic equation follow from the state diagram, the reverse is also true. That is, given the arithmetic equation, it is an easy matter to construct the state diagram.

In the next example, the least amount that can be measured is not 1. Let the decanters have sizes 15 and 99. Before reading on, can you see why it is impossible to obtain exactly one unit of water? An equation can be obtained for any sequence of moves. Such an equation is of the form

$$15x + 99y = z$$

where exactly one of the integers x and y is negative, and z is the amount obtained. Now notice that the left side is a multiple of 3, so the right side must be also. Thus the least positive amount that can be measured is a multiple of 3. One can also reason this as follows: each fill adds a multiple of 3 units of water to the total amount on hand, each pour leaves the total number unchanged, and each dump removes a multiple of three units from the total, so the amount on hand at each stage is a multiple of 3.

In general, when a and b are integers, the least amount that can be measured is the greatest common divisor of the two decanter sizes, and the Euclidean algorithm, as explained below, tells us how to proceed. Suppose $c = GCD(a, b)$. The Euclidean algorithm yields a solution to

$$c = ax + by$$

where x and y are integers exactly one of which is positive and, except in trivial cases, the other is negative. For convenience, we assume x is positive. Then the solution to the decanting problem is to fill the a capacity decanter x times, repeatedly pouring its contents into the b unit decanter. The b unit decanter will be dumped y times, so the total liquid on hand at the end is the difference $ax - by = c$.

Solutions to Additional Problems 161

Let's look at another specific example. Again we use the Euclidean Algorithm to solve the decanting problem. There are two stages. The first stage is a sequence of divisions. The second is a sequence of 'unwindings.' For this example, let the decanter sizes be $a = 257$ and $b = 341$. Use the division algorithm to get $341 = 1 \cdot 257 + 84$. Then divide 257 by 84 to get $q = 3$ and $r = 5$. That is, $257 = 3 \cdot 84 + 5$. Continue dividing until the dividend is less than the divisor. Thus 84 divided by 5 yields $84 = 16 \cdot 5 + 4$. Finally, divide 5 by 4 to get $5 = 1 \cdot 4 + 1$. This completes the first stage. Now to unwind, start with the final division scheme writing $1 = 5 - 1 \cdot 4$. Then replace the 4 with $84 - 16 \cdot 5$ to get $1 = 5 - 1(84 - 16 \cdot 5)$. This is equivalent to $1 = 17 \cdot 5 - 1 \cdot 84$. Check this to be sure. Then replace 5 with $257 - 3 \cdot 84$ to get

$$1 = 17 \cdot (257 - 3 \cdot 84) - 1 \cdot 84,$$

i.e., $1 = 17 \cdot 257 - 52 \cdot 84$. Finally, replace 84 with $341 - 257$ to get $1 = 17 \cdot 257 - 52(341 - 257)$, which we can rewrite as

$$1 = 69 \cdot 257 - 52 \cdot 341.$$

Thus, the solution to the decanting problem is to measure out 1 unit of liquid by filling the 257 unit decanter 69 times, repeatedly pouring its contents into the 341 unit decanter, and, in the process, dumping out the 341 unit decanter 52 times.

Now back to the bill splitting problem. Imagine that we have two decanters with capacities $a = 199$ and $b = 98$. Notice that $GCD(199, 98) = 1$. As we did above, we can use the Euclidean algorithm to find numbers x and y satisfying $199x + 98y = 1$ where exactly one of the numbers x, y is negative. We do this by dividing repeatedly. First, 98 into 199 yields $199 = 2 \cdot 98 + 3$, Then 3 into 98 yields $98 = 32 \cdot 3 + 2$ and finally we can write $1 = 3 - 2$. Next we go to the unwinding stage.

$$\begin{aligned}
1 &= 3 - 2 \\
&= 3 - (98 - 32 \cdot 3) \\
&= 3 - 98 + 32 \cdot 3 \\
&= 33 \cdot 3 - 1 \cdot 98 \\
&= 33(199 - 2 \cdot 98) - 98 \\
&= 33 \cdot 199 - 66 \cdot 98 - 98 \\
&= 33 \cdot 199 - 67 \cdot 98.
\end{aligned}$$

Thus, we have the values $s = 67$ and $t = 33$. Indeed, $2 \cdot 33.67 - 67.33 = 0.01$.

2. The 7-11 problem

Solution. The four prices are \$1.25, \$1.20, \$1.50, and \$3.16. To see how to get these numbers, let $\underline{x}$, $\underline{y}$, $\underline{u}$, and $\underline{v}$ denote the four prices, in dollars. Then $\underline{xyuv} = 7.11$ and $\underline{x} + \underline{y} + \underline{u} + \underline{v} = 7.11$. To eliminate the fractional part, multiply each of the unknowns and rename to get $x = 100\underline{x}$, $y = 100\underline{y}$, $u = 100\underline{u}$, and $v = 100\underline{v}$. Thus we have $xyuv = 10^8 \cdot 7.11$ and $x + y + u + v = 711$. Factor the former to get $xyuv = 711 \cdot 10^6 = 2^6 \cdot 3^2 \cdot 5^6 \cdot 79$. It follows that exactly one of x, y, u, v must be a multiple of 79. For convenience, let's say it is v. Then v is one of the seven numbers 79, 158, 237, 316, 395, 474, or 612. We argue that the last three of these are too big. For example, suppose $v = 395$. Then $xyu = 711 \cdot 10^6 \div 5 \cdot 79 = 18 \cdot 10^5$. Now the least possible sum $x + y + u$ occurs when $x = y = u = \sqrt[3]{18 \cdot 10^5} \approx 121.6$, in which case $x + y + u + v > 711$. A similar argument works for $v = 474$ and $v = 612$ as well. Next consider $v = 316$. In this case, $xyu = 711 \cdot 10^6 \div 316 = 2^4 \cdot 3^2 \cdot 5^6$ and $x + y + u = 711 - 316 = 395$. Note that $\sqrt[3]{2^4 3^2 5^6} = 50\sqrt[3]{18} > 125$, so the sum $x + y + u$ must be at least $3 \cdot 125 = 375$. Therefore we try to minimize $x + y + u$ subject to $xyu = 2^4 3^2 5^6$. This occurs when we choose $x, y,$ and u as close together as possible. Hence, let $x = 5^3 = 125$, $y = 2^3 \cdot 3 \cdot 5 = 120$ and $u = 2 \cdot 3 \cdot 5^2 = 150$. Amazingly, this works. The other three possible values of v, 79, 158 and 237 can be eliminated, but the work is rather tedious. Thus the four prices are $\underline{x} = \$1.25$, $\underline{y} = \$1.20$, $\underline{u} = \$1.50$ and $\underline{v} = \$3.16$. In the three-item problem, we have the following: 6.00 (1, 2, 3); 8.25 (.75, 2, 5.5); 9.00 (.5, 4, 4.5); 10.80 (.4, 5, 5.4).

3. Longest path problem

Solution. We think of the grid as a graph, with 18 vertices, and 25 edges. The vertices have *degrees* 2 and 3. There are four vertices of degree 2 and 14 of degree 3. The edges also come in two types, those with length 2 (there are 7 of these) and those with length 1 (there are 18 of these). Since we are starting at A, only one of the edges adjacent to A can be part of the path. The same is true for B. At all the other vertices, we can and must use exactly two edges. At the corners we have only two edges, but at all the other vertices, we must choose one of the two incident edges. It is possible to choose a path so that the edges that are not used all have length 1. If every vertex belongs to the path, then exactly 17 edges can belong to the path, which means 8 edges do not belong. If all these edges have length 1, then the length of the path must be maximal, $L = 2 \cdot 7 + 18 \cdot 1 - 8 \cdot 1 = 14 + 18 - 8 = 24$. Note that the path below from A to B has length 24.

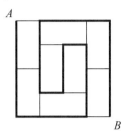

4. The Wizards

Solution. Let's call a positive integer *ambiguous* if there are two different partitions of equal size of it into summands whose products are the same. For example, 12 is ambiguous because $12 = 2+2+2+6 = 1+3+4+4$ and $2 \cdot 2 \cdot 2 \cdot 6 = 1 \cdot 3 \cdot 4 \cdot 4 = 48$. A number is called *doubly ambiguous* if there are two pairs of partitions each of which has the same product, and these two products are different. For example, 13 is doubly ambiguous since $13 = 1+6+6 = 2+2+9$ and $1 \cdot 6 \cdot 6 = 2 \cdot 2 \cdot 9 = 36$ and at the same time $13 = 1+1+3+4+4 = 1+2+2+2+6$ and $1 \cdot 1 \cdot 3 \cdot 4 \cdot 4 = 1 \cdot 2 \cdot 2 \cdot 2 \cdot 6 = 48$, while 12 is not doubly ambiguous. Now the first wizard's response to the second wizard's comment is equivalent to 'the bus number is ambiguous.' Note that if n is ambiguous, then so is $n+1$ and if n is doubly ambiguous, then $n+1$ is also doubly ambiguous. An examination of all partitions of 11 shows that 11 is not ambiguous. Thus every integer $n \geq 12$ is ambiguous, and every integer $n \geq 13$ is doubly ambiguous. The second wizard's comment that he knew the age of the first wizard implies that the bus number is not doubly ambiguous. Thus the bus number must be 12.

5. A Pebbling Problem

Solution. Before we can solve this problem, we need some preliminary material. Consider the following problem. Find a pair of integers m and n such that

$$m/n = 0.3636363\ldots.$$

You've seen problems like this in your algebra course, reinforcing the idea that every rational number has a repeating decimal representation. To solve it let $x = 0.36363636\ldots$ (which we can write as $0.\overline{36}$) in which case $100x = 36.36363636\ldots$. Subtract the former equation from the latter to get $99x = 36$, which leads to $m = 4$ and $n = 11$. This is really the same type of problem (in disguised form) as the following. Find the value of the geometric series

$$\frac{2}{3} + \frac{2}{9} + \frac{2}{27} + \cdots.$$

Again, give the answer a name. Let $S = \frac{2}{3} + \frac{2}{9} + \frac{2}{27} + \cdots$. Then $3S = 3 \cdot \frac{2}{3} + 3 \cdot \frac{2}{9} + 3 \cdot \frac{2}{27} + \cdots = 2 + \frac{2}{3} + \frac{2}{9} + \frac{2}{27} + \cdots = 2 + S$. Subtract just as before to get $2S = 2$ or $S = 1$. Try this with the geometric series $.9 + .09 + .009 + \cdots = 0.\overline{9}$ for a result that may surprise you. We can now solve the general problem: find $S = a + ar + ar^2 + \cdots$ where a and r are given and $|r| < 1$. Multiply both sides by r to get $rS = ar + ar^2 + ar^3 + \cdots$ and subtract to find that $S - rS = a$, in which case $S = \frac{a}{1-r}$.

Now back to the pebbling problem. Assign each square a value $v(i, j)$, $i = 0, 1, 2, \ldots, j = 0, 1, 2, \ldots$ as follows: $v(i, j) = 2^{-(i+j)}$. Thus we have values as shown in the grid:

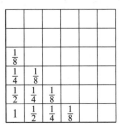

Next let us define the value of a *position* of the puzzle. Each move, that is, each replacement of a counter by two counters, results in a new position of the puzzle. The *value* of a position of the puzzle is the sum of the values of the squares occupied in that position. The value of the initial position is 1. We compute the value of various positions of the puzzle. The position has value $v = 1/2 + 1/2 = 1$, while has value $v = 1/2 + 1/4 + 1/4 = 1$. Is it clear that the value of each position is obtainable from the value of the previous position by removing $1/2^{-n}$ from the sum and replacing it by $1/2^{-n-1} + 1/2^{-n-1}$, thus, not changing the value. Now, if there is a position of the puzzle where all the counters are outside of the shaded region, such a position must have the value 1. However, let us compute the sum of the values of *all* the squares outside the shaded region. We'll do this column by column. The values of the squares in the first column are $1/16 + 1/32 + 1/64 + \cdots = 1/8$, so let us enter the number 1/8 for bookkeeping purposes.

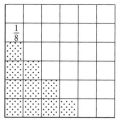

Notice next that second, third, fourth and fifth columns have the same sum, 1/8. Now our diagram looks like

The sixth, seventh, etc columns have sum $1/16, 1/32, 1/64$, etc. Now our diagram looks like

We can add the values of the bottom row in the usual way to get 1/4. Hence the value of the entire first quadrant, minus the 10 shaded squares is $1/8 + 1/8 + 1/8 + 1/8 + 1/4 = 3/4$, so it is impossible to move all the counters outside the shaded area.

6. Thirty Digits

Solution. Let N denote any 30-digit number of the form $2^i 5^j 7^k$. Suppose that every digit appears exactly three times in the decimal representation of N. Then the sum of the digits is $3(0 + 1 + 2 + \cdots + 9) = 3 \cdot 45 = 135$ which is divisible by 9. This implies that N itself is divisible by 9. But the Fundamental Theorem of Arithmetic asserts that N has only one factorization into primes. The prime factors of N are 2, 5, and 7, but not 3, so we have a contradiction.

7. The Whispered Number

Solution. Version A. P's statement implies that p is odd. Q's statement implies that q is odd and also that $q \geq 7$. There are just six triplets (p, q, r) that satisfy all three statements (a) $p + q + r = 14$, (b) p is odd and (c) $q \geq 7$ and odd. They are $(1, 7, 6), (1, 9, 4), (1, 11, 2), (3, 7, 4), (3, 9, 2)$, and $(5, 7, 2)$. Since student R reasons perfectly, he can establish that only the six triples listed above are possible. In order to make his statement, r must be 6. Thus the product is $1 \times 7 \times 6 = 42$.

Version B. P's statement implies that p is odd. Q's statement implies that q cannot be 1, 3, or 5 because he could not know that his number is different from P's. If $q = 4, 8$ or 12, Q could not know that p is different from r, because we could have $p = r = 5, p = r = 3$, and $p = r = 1$ respectively. Also, q is not 7, 9, or 11 because if it were, Q would have known already that all three numbers were different. Thus $q = 2, 6,$ or 10. There are just 10 triplets (p, q, r) that satisfy all three statements (a) $p + q + r = 14$, (b) p is odd and (c) $q \in \{2, 6, 10\}$. They are $(1, 2, 11), (1, 6, 7), (1, 10, 3), (3, 2, 9), (3, 6, 5), (3, 10, 1), (5, 2, 7)$, $(5, 6, 3), (7, 2, 5)$, and $(7, 6, 1)$. Since student R reasons perfectly, he can establish that only these triples listed above are possible. In order to make his statement, r must be one of the uniquely mentioned values, 9 or 11. The triplet $(1, 2, 11)$ does not work because R would have known from P's statement that $p = 1$ and $q = 2$. Hence the only possible triplet is $(3, 2, 9)$ and the product is $3 \times 2 \times 9 = 54$.

8. The Staircase Problem

Solution. The answer is that there are $\binom{n+3}{4}$ rectangular regions bounded by the gridlines of an n-staircase. (This solution is due to James Rudzinski.) Let G_n denote the total number of rectangular regions in an n-staircase, $n \geq 1$. Also let the unit square that is the intersection of the longest column and the longest row of the n-staircase be called square A. First, note that if we remove either the longest column or the longest row from the n-staircase, we would get an $(n-1)$-staircase. The number of rectangular regions in either of these subsections of the n-staircase would be G_{n-1}. If we count the rectangular regions in both subsections, then we will get a double-count of all of the rectangular regions in the intersection of the two $(n-1)$-staircase subsections. Also notice that the intersection of the two $(n-1)$-staircases is what is left after both the longest column and the longest row are removed from the n-staircase, which is simply an $(n-2)$-staircase. So far we have counted $2G_{n-1} - G_{n-2}$ rectangular subregions, which includes every rectangular subregion except those subregions that contain square A. It remains only to count the number of subregions that contain square A. Note that in this case one simply needs to choose any other intersection not on the outer edges of the longest column or longest row as the opposite corner to form another rectangular region. The number of choices for such an intersection is a triangular number, since the number of choices starts with one choice in the first column, and the number of choices per column goes up in increasing intervals of one, up to the last column, which has n choices. Thus the number of rectangular regions containing square A is $n(n+1)/2 = \binom{n+1}{2} = \binom{n+1}{n-1}$. Thus G_n is equal to $2 \cdot G_{n-1} - G_{n-2} + \binom{n+1}{2}$. Now, noting that $G_1 = 1 = \binom{4}{4}$ and $G_2 = 5 = \binom{5}{4}$ (by observation), we get

$$G_3 = 2\binom{5}{4} - \binom{4}{4} + \binom{4}{2} = \binom{5}{4} + \binom{4}{3} + \binom{4}{2} = \binom{5}{4} + \binom{5}{3} = \binom{6}{4}.$$

If we now inductively assume that $G_k = \binom{k+3}{4}$ for some positive integer k, we get

$$G_{k+1} = 2G_k - G_{k-1} + \binom{k+2}{2} = 2\binom{k+3}{4} - \binom{k+2}{4} + \binom{k+2}{2}$$
$$= \binom{k+3}{4} + \binom{k+2}{3} + \binom{k+2}{2} = \binom{k+3}{4} + \binom{k+3}{3}$$
$$= \binom{k+4}{4} = \binom{(k+1)+3}{4}$$

and this completes the inductive proof. Thus we finally get $G_n = \binom{n+3}{4}$ for $n \geq 1$.

Alternatively, the case for n can be reduced to the $n-1$ case in a third way: by stripping the top square off of each column. Along with the two reductions mentioned above, one is led to the recursion relationship $G_n = 3G_{n-1} - 3G_{n-2} + G_{n-3} + n$. The first three terms are the usual inclusion/exclusion argument. The last accounts for the uncounted squares, those that include the bottom right square and a square from the top of a column ... there are n of these. The desired formula follows.

9. Piles of Stones

Solution. Note that both 5 and $49 + 51$ are multiples of 5, so if the first move is to add the two big piles, then every pile after that will have a multiple of 5 stones. On the other hand, both $5 + 49$ and 51 are multiples of 3, so with this first move, all the piles will have a multiple of 3 stones after that. Finally, $5 + 51 = 56$ which, like 49, is a multiple of 7. Once all the pile sizes are multiples of the same odd prime, that situation persists. Thus, no piles of size 1 are possible.

10. Flipping Pennies

Solution. Suppose there are k heads on the table. Take any k coins and make them into a group. Turn them all over. Suppose that among the k coins selected, there are h heads. Then the other group has $k - h$ heads. Once all the coins in the group with k are turned over, there will be $k - h$ heads in that group as well.

11. Back to Front

Solution. First we describe a six-move sequence that works. Then we prove that no shorter sequence will do.

$$11, 10, 9, 8, 7, 6, 5, 4, 3, (2, 1);$$
$$(11, 10, 9, 8, 2), 1, 7, 6, 5, 4, 3;$$
$$1, 7, 6, 5, (4, 11), 10, 9, 8, 2, 3;$$
$$1, 7, 6, (5, 10), 9, 8, 2, 3, 4, 11;$$
$$1, 7, (6, 9), 8, 2, 3, 4, 5, 10, 11;$$
$$1, (7, 8), 2, 3, 4, 5, 6, 9, 10, 11;$$
$$1, 2, 3, 4, 5, 6, 7, 8, 9, 10, 11.$$

On the other hand, six moves is the best possible. To see this, call two consecutive cards a, b a *disorder* if $a > b$. There are 10 disorders initially. The first and last moves can remove at most one disorder, and any other move can diminish the number of disorders by at most two (if three are removed, then one is created). So the fewest number of moves needed to remove all the disorders is $1 + 1 + 8/2 = 6$.

12. Face Painting

Solution. Suppose the cube is $n \times n \times n$, and the number of faces painted red is i. Then the number of faces painted black is $6 - i$. Of course, $i \geq 6 - i$, so $i = 3, 4, 5$, or 6. We can eliminate $i = 3$ quickly since this leads to the same number of cubes with some red paint as those with black paint.

Solutions to Additional Problems 169

Also, if $i = 6$, then all the surface cubes have some red paint and none have black paint. The number of cubes in the interior of the $n \times n \times n$ cube is $(n-2)^3$. Thus we have the equation $n^3 - (n-2)^3 = 200$, and it is easy to see that this equation has no integer solutions. We can also eliminate the case $i = 5$. In this case, there are $6n^2 - 12n + 8 - (n-2)^2$ cubes with some red paint and n^2 cubes with some black paint. Simplifying and setting the difference equal to 200 yields $4(n-1)^2 = 200$, which fails for each integer value of n. Thus, we conclude that $i = 4$.

Here there are two cases. Either the two black faces are adjacent or opposite. In case they are opposite, the number of faces with some red paint is $n^3 - (n-2)^3 - 2(n-2)^2$ and the number of faces with some black paint is $2n^2$. So the question is, is there an integer n such that $n^3 - (n-2)^3 - 2(n-2)^2 - 2n^2 = 200$? The equation simplifies to $n(n-2) = 100$, which is clearly not solvable over the integers.

In case the two black faces are adjacent, the number of faces with some red paint is $n^3 - (n-2)^3 - 2(n-2)^2 - (n-2)$ and the number of faces with some black paint is $2n^2 - n$. The difference is $2n^2 - 4n - 2 = 2(n-1)^2$, which has the value 200 when $n = 11$. Thus the cube is $11 \times 11 \times 11$ and there are $(11-2)^3 = 729$ cubes without any paint.

13. High Slopes

Solution. The answer is 870. (The following argument is due to Anders Kaseorg.) For convenience we first solve the minimum problem. Color the squares black and white as on a chessboard, such that every odd number goes on a white square. If every number on the white diagonal is smaller than 59, then the 21 numbers $59, 61, \ldots, 99$ must be on the same side of that diagonal—but there are only 20 white squares on each side. Therefore, the sum of the white diagonal is at least $1 + 3 + \cdots + 17 + 59 = 140$. We get a similar contradiction if every number on the black diagonal is smaller than 60, so the sum of the black diagonal is at least $2 + 4 + \cdots + 18 + 60$. Thus the smallest possible sum of the numbers on the diagonal is 140.

To see that this value can be achieved, we start on the diagonal and stay on it, including as many of the first few numbers as possible. If we start in the upper left corner, note that only odd numbers will appear on the diagonal. We can zigzag our way down the diagonal, putting the numbers $1, 3, 5, 7, 9, 11, 13, 15$, and 17 on the diagonal as shown. At this stage we cannot hope to put 19 on the diagonal because we would then not have access to both the squares above the diagonal and below the diagonal. We can, however use up all the squares above the diagonal, then use the final diagonal square, and finally complete the arrangement by listing all the rest of the numbers in the square below the diagonal. This results in putting the number 59 in the bottom right corner, so the sum of the entries on the diagonal is $1 + 3 + \cdots + 17 + 59 = 140$. In general, for a $2n \times 2n$ grid, the smallest sum of numbers along the diagonal is the sum of the first $2n - 1$ positive odd integers plus the number $2n^2 + 2n - 1$. Hence the smallest possible sum is $(2n - 1)^2 + 2n^2 + 2n - 1 = 6n^2 - 2n$. To get the maximum possible sum, we can replace each diagonal number k by $101-k$. The result is $8n^3 - 6n^2 + 4n$, which for $n = 5$ gives the value 870.

1	2						
	3						
	4	5	6				
			7				
			8	9	10		
					11		
					12	13	14	19
							15	18
							16	17

14. An Odd End

Solution. The product of the ten numbers is

$$P(n) = n \cdot (n+1) \cdot (n+2) \cdots (n+9).$$

Consider all the prime factors of $P(n)$. For the last non-zero digit to be odd, all the factors of 2 must be taken care of among the zeros: the number x of factors of 2 must be less than or equal to the number of factors of 5. Two of the ten numbers will have at least one factor of 5, with at most one of them having a factor of more than one 5. So if a multiple of 5^u, say, is in the list of ten, the number of factors of 5 is $y = u + 1$. So we need $x \leq y$. But the ten numbers include five evens, at least two of which are multiples of four, and at least one of which is a multiple of eight. So the minimum possible value of x is $1 + 1 + 1 + 2 + 3 = 8$. So the minimum value of y is 8. Thus we'll look for a list of ten numbers that include a multiple of $5^7 = 78125$, starting at the lowest possible. $P(78116)$ has eight 5's but more than eight 2's. However, $P(78117)$ has eight 5's and eight 2's, as required. So the least starting number is 78117. The rightmost nonzero digit of $P(78117)$ is the digit 7.

15. Prime Leaps

Solution. The longest list can contain at most two from any eight consecutive numbers in $0, 1, 2, 3, 4, \ldots$. This is because if k is in the list, then $k + 2, k + 3, k + 5$, and $k + 7$ cannot be. And at most one of $k + 1, k + 4$, and $k + 6$ can be in the list. So there is no hope of choosing 501 numbers all differing by non-primes (because the third is at least 8, the fifth is at least 16, ..., and the 501st is at least 2000). We'll now show how to choose 500 numbers and we'll show that the choice is unique. Clearly the list $0, 4, 8, 12, 16, \ldots, 1996$ contains 500 numbers and any two differ by a multiple of 4 (and so no difference is prime). Is there any other way of getting 500 numbers? Suppose we choose 0 first. Then the third choice must be at most 8 (because we cannot fit 498 choices in $9, 10, \ldots, 1996$). Similarly, the fifth choice must be at most 16, etc. We may as well choose 0 first (and the next must then be 1, 4, or 6). If we have chosen 1 also, then 2 through 8 must be ruled out, but that means the third is greater than 8. So we cannot choose 1 as the second. If we choose 6 next, then 7 and 8 are ruled out. In these cases, the third number is greater than 8. Thus the second number must be 4. Now we can start the argument again to deduce that the third choice must be 8, etc. So the answer is $0, 4, 8, 12, \ldots, 1996$; that is, all the years divisible by 4.

16. Multiple Quotients

Solution. The numbers are precisely $(1abc\ldots)/(2uvw\ldots)$ where each of the numbers $1, 2, 3, 4, \ldots, 11$ appears once. Another way to write these numbers is as $11!/(2uvw\ldots)^2$ where $u, v, w, \ldots$, are among $3, 4, \ldots, 11$. The largest integer is $11!/2 = 9979200$. Since $11! = 11 \cdot 7 \cdot 5 \cdot 5 \cdot 3 \cdot 3 \cdot 3 \cdot 2 \cdot 2 \cdot 2 \cdot 2 \cdot 2 \cdot 2 \cdot 2 = 11 \cdot 7 \cdot (2 \cdot 5 \cdot 8 \cdot 9)^2$, it follows that the smallest integer is $11!/(2 \cdot 5 \cdot 8 \cdot 9)^2 = 77$. The quotient in question is $9979200/77 = 129600$.

17. Unsquare Party

Solution. A relation R on a set S is a set of ordered pairs (x, y), where both x and y belong to S. If R is a relation on a set S, instead of writing $(x, y) \in R$, we write xRy, and say 'x is related to y.' Define a relation

$\sim$ on the set $S = \{1, 2, 3, 4, \ldots, 100\}$ as follows:

For any $x, y \in S$, $x \sim y$ if $x \cdot y$ is a perfect square.

Thus, for example $1 \sim 1$, $1 \sim 4$, $2 \sim 8$, and $3 \sim 12$. Next note that $\sim$ satisfies

1. $x \sim x$ for all $x \in S$ (*reflexive* property),
2. $x \sim y$ implies $y \sim x$ for all $x, y \in S$ (*symmetry* property), and
3. $x \sim y$ and $y \sim z$ implies $x \sim z$ for all $x, y, z \in S$ (*transitive* property).

Relations that satisfy all three properties above are called *equivalence relations*. If R is any relation on a set S and $x \in S$, define the *cell* of x, denoted $[x]$ as follows:

$$[x] = \{y \mid xRy\}.$$

In the case when R is an equivalence relation, the cells of the members of S have a special property: two cells, $[x]$ and $[y]$ are either identical or they are disjoint. In other words, $[x] = [y]$ or $[x] \cap [y] = \emptyset$. We leave the proof of this fact to the reader.

Our relation $\sim$ is an equivalence relation. To see this, note that (a) $x \sim x$ for all x because $x \cdot x$ is a perfect square; (b) if $x \sim y$, then $y \sim x$ because $x \cdot y = y \cdot x$; and (c) if $x \sim y$ and $y \sim z$, then $x \cdot y$ and $y \cdot z$ are both perfect squares. It follows that $x \cdot z = xy^2z \div y^2$ is also a perfect square, so $x \sim z$.

Finally, to answer the question, note that Ashley can have at most one relative whose age belongs to a given cell. The maximum number of relatives Ashley could have is the number of cells. To count the cells, we can list the members of $[1] = \{1, 4, 9, 16, 25, 36, 49, 64, 81, 100\}$. Then find the smallest member of S not in one of the cells listed, and find its cell. Thus $[2] = \{2, 8, 18, 32, 50, 72, 98\}$. Continuing in this way, we find that there are 61 cells. Alternatively, note that the smallest member of each cell is square free. Here we count 1 as square free. Furthermore, each cell can contain at most one square free number because the product of distinct square free numbers cannot be a square. We can count the square free numbers less than or equal to 100 using the Inclusion/Exclusion Principle:

$$100 - \left\lfloor \frac{100}{2^2} \right\rfloor - \left\lfloor \frac{100}{3^2} \right\rfloor - \left\lfloor \frac{100}{5^2} \right\rfloor - \left\lfloor \frac{100}{7^2} \right\rfloor + \left\lfloor \frac{100}{2^2 3^2} \right\rfloor + \left\lfloor \frac{100}{2^2 5^2} \right\rfloor = 61$$

Solutions to Additional Problems

18. Allison's Coin Machine

Solution. The answer is no. Notice that the exchange *nickel* → 5 *pennies* leaves the value unchanged. The exchange *penny* → 5 *nickels* increases the value by 24 cents. Therefore, Allison's coins always have total value $1 + 24n$ for some integer n, an odd number. But if she had the same number of pennies and nickels, the value of her coins would be an even number of cents.

19. Chameleons

Solution. No, it is not possible. To see this, let (x, y, z) denote the number of grey, brown, and crimson chameleons respectively. If two different color chameleons meet, one of the three changes takes place: $(x, y, z) \mapsto (x - 1, y - 1, z + 2)$; $(x, y, z) \mapsto (x - 1, y + 2, z - 1)$; or $(x, y, z) \mapsto (x + 2, y - 1, z - 1)$. Note that the difference between the number of grey chameleons and brown chameleons always changes by zero or 3. This is also true about the difference between the number of grey and crimson chameleons as well as the difference between the number of brown and crimson chameleons. Since no pair of these numbers initially differ by a multiple of 3, they can never be made the same.

20. Repeated Arithmetic

Solution. No sequence of operations can return the set of numbers to the original set because the sum of the squares of the two new numbers is larger than the sum of the squares of the original numbers.

21. Double or Add Seven

Solution. The set S cannot have a multiple of 7 because none of the productions (a), (b), or (c) can produce one. Also, it is not hard to see using modular arithmetic that any non-multiple of 7 larger than 1000 does belong to S. The smallest multiple of 7 larger than 2004 is 2009.

22. Double or Subtract Twelve

Solution. The function f takes multiples of three to themselves so there is no k for which $f^k(3n) = 16$. On the other hand, if n is not a multiple of 3, then $n, f(n), f^2(n), \ldots$ is eventually larger than 25, and repeatedly subtracting 12 leaves one of the numbers 13, 14, 16, 17, 19, 20, 22, or 23. It is not hard to check that each of these has an iterate of 16. So $1000 - 333 = 667$ of the first thousand positive integers have 16 as an iterate.

23. Counting Transitive Relations

Solution. The answer is 171. First note that the set of 512 relations can be partitioned into 10 classes by their number of ordered pairs. The number in each cell is an entry in the 10th row of Pascal's triangle. In the matrix below, each row represents the number of relations with 0, 1, 2, or 3 *loops* and the given number of ordered pairs. A loop is an ordered pair of the form (x, x). Thus the loops are the ordered pairs, $L = \{(a, a), (b, b), (c, c)\}$ and the nonloops are $N = \{(a, b), (a, c), (b, a), (b, c), (c, a), (c, b)\}$. The boldface 45, for example, in row 2, position 6 means there are 45 relations on $\{a, b, c\}$ with six edges, two of which are loops. To compute this number note that there are $\binom{3}{2} = 3$ ways to pick two of the three loops, and $\binom{6}{4} = 15$ ways to pick the four nonloops, the product of which is 45.

loops \ edges	0	1	2	3	4	5	6	7	8	9
0	1	6	15	20	15	6	1	0	0	0
1	0	3	18	45	60	45	18	3	0	0
2	0	0	3	18	45	60	**45**	18	3	0
3	0	0	0	1	6	15	20	15	6	1
totals	1	9	36	84	126	126	84	36	9	1

Note that totals in the bottom row represent the number of relations with a given number(the column number) of ordered pairs.

The matrix below gives the number of transitive relations in each of these positions.

loops \ edges	0	1	2	3	4	5	6	7	8	9
0	1	6	6	6	0	0	0	0	0	0
1	0	3	18	18	18	0	0	0	0	0
2	0	0	3	18	21	18	**6**	0	0	0
3	0	0	0	1	6	9	6	6	0	1
totals	1	9	27	43	45	27	12	6	0	1

The bold 6 means, for example, that of the 45 relations with six edges, two of which are loops, there are six transitive ones. These all look basically the same. Pick two members of S (from the three choices), say a and b. Include in R the four edges $(a, a), (a, b), (b, a)$, and (b, b). Then add either (b, c) and (a, c) or (c, a) and (c, b) (two choices), for a total of $3 \cdot 2 = 6$ such relations. Finally the sum $1+9+27+43+45+27+12+6+1 = 171$ is the number of transitive relations on the set $\{a, b, c\}$.

Classification

Many AMC12 and AMC10 problems combine several diverse areas of mathematics. Therefore, their classification is more difficult. Several problems appear multiple times in the listing below. They are classified mainly by the type of mathematics that could be used in the solution.

The general classifications are:
- Algebra, including analytic geometry, function, and logarithms.
- Complex numbers.
- Discrete Mathematics, including counting problems, logic and discrete probability.
- Geometry, including three-dimensional geometry.
- Number theory, including divisibility, representation, and modular arithmetic.
- Statistics
- Trigonometry

The references to the problems gives the number of the AMC 12 followed by number of the problem. For example, 47-n refers to the nth problem on the 47th AMC 12. The exception to this is the references to the AMC 10 problems, which are denoted 0-n and 1-n. Finally, the notation An refers to the problem from year $1900 + n$ on the Anniversary Test.

Algebra

Absolute Value: 48-22, 48-28, 0-20, 50-22, 51-5 (1-9), A77

Arithmetic: 46-3, 46-13, 47-1, 47-2, 47-3, 47-5, 47-6, 48-1, 48-7, 49-2, 49-3, 49-5, 50-1, 50-4

Approximation: 46-5, 46-16
Binomial Theorem: A61
Calendar: 0-13, 50-8, 51-18 (1-25)
Circle, Equation of: 47-20, 47-25, 49-23
Complete the Square: 49-23
Conjugate of Radical Expression: A50
Coordinate Geometry: A63, A64, A89, 47-17, 47-28, 49-19, 49-29, 50-22 (0-20), 51-10, 51-16, 51-21
Cubic Equation, Cubic Function: 0-19
Decimal Arithmetic: 47-7
Defined Operation: A62
Distance Formula: 46-21, 47-27 (A96), 47-28, 48-3
Distance, Velocity, and Time: 46-7, 47-13, 49-21, A68
Distributive Law: A62
Equations, Simultaneous Nonlinear: A76
Exponents: 47-6, 51-2 (1-2), 49-5, A69, A84
Factorization into Primes: 46-24, 46-29
Factoring: 50-30, 51-12, A67
Floor (greatest Integer) Function: A70
Fractional Powers (radicals): 49-7
Fractions: 47-5, 49-2, 50-3, 0-1, 51-11 (1-15), 51-20
Function: Composite: 51-15 (1-24)
Function, Graph: 0-19, 50-22 (0-20), 51-20, A92
Function, Periodic: 48-27
Functional Equation: 49-17, A79
Lattice Points: 49-29, A89
Line:
  Equation of: 46-12, 48-12, A64
  Graph of: A80, A92
  Intercepts: 47-7
  Slope of: 48-12
  y-intercept of: 48-12
Linear Equation: 47-2, 47-7, 48-8, 50-8, 0-13, 1-8, A55
Linear Equations, Simultaneous: 46-10, 49-25, 50-27, 50-28, 1-4, A68, A86

Classification

Logarithms:
  Base ten: 46-5, 50-18 (A99), 51-23, A54, A85
  Base Variable: 51-7
  Other: 46-28, 47-8, 47-13, 47-23, 48-17, 49-12, 49-22 (A98), 50-28
Nonlinear Equations, Simultaneous: 46-28, 47-8, 47-13, 47-23, 50-22
Palindrome: 50-9
Percent: 46-4, 46-16, 48-4, 49-9, 50-5, 0-2, 51-3 (1-3)
Polynomials:
  Properties of Coefficients: 46-14, 51-22, A61
  Quotient of: 50-17
  Relation between Zeros: 50-12, 51-15 (1-24)
Powers: 48-21
Quadratic, Equation: 49-14
Quadratic Formula, Discriminant: 47-25, 48-19
Quadratic Function, Minimum Value: 0-19
Rational and Irrational Numbers: A50, A74
Ratio and Proportion: 48-13, 49-15, 0-9, A51
Sequence:
  Arithmetic: 48-20, 51-14 (1-23), 51-16, A60
  Geometric: 50-13
  Other: 47-24, 48-6, 48-13, 50-20, 0-5, 51-4, A89
Square Roots: 46-2
Symmetry: 51-20
Ternary Operation: 49-4
Units, Relative: 46-4

Complex numbers

DeMoivre's Theorem: A81

Discrete Mathematics

Counting:
  Adjacencies: 49-1
  Combinations: 46-29, 47-10, 47-22, 47-26, 48-24, 48-30, 50-24, 0-15
  Lattice Points: 46-30 (A95), 47-27 (A96), A89
  Orderings: 49-20, A87
  Permutations: 49-24, A87
  Polygons: 46-9, 46-23, 49-19

n-Digit Integers: 46-11, 48-30, 50-1
Other: 47-29, 48-28, 49-11, 50-14, 0-17, 51-25, 1-8, 1-13, A65
Factorial: 47-3, 49-30, 50-25
Graph Theory: 46-15, 49-1
Invariant: 1-17, A91
Inclusion-Exclusion Principle: 48-28, 49-11, 49-24, 49-27, 50-14, A65, A83
Logic: 49-20, 50-2, 50-10, 0-21, 1-21, A75, A78
Matrix: 48-16
Orderings around a Circle: 47-22
Paths on a Grid: A89,
Pattern Finding: A91, A93
Pattern, Geometric: 46-9, 51-8 (1-12), A89
Pigeon Hole Principle: 48-16
Probability:
    Continuous, Geometric: 0-16, 50-29
    Dice, Fair: 47-16, 48-10
    Finite: 46-20, 47-22, 47-26, A83, A85
Recursive Sequence: 46-27, 47-12

Geometry

Angles:
    Bisection: 49-28, 51-17, 51-19, 1-5
    in Polygon: 46-18, 50-7, 50-26, 0-24
Area:
    of Circle: 46-26, 49-16, 0-22, 0-24
    of Hexagon: 47-19, 49-15, 50-23
    of Pentagon: 46-22
    of Polygon: 51-8
    of Quadrilateral: 48-9
    of Rectangle: 47-11, 49-9, 49-29, A51
    of Rhombus: 50-16
    of Semicircle: 0-24
    of Trapezoid: 49-8, 1-5
    of Triangle: 46-10, 47-15, 49-19, 0-24, 51-21, A57
        by Determinants: 46-10
        Equilateral: 46-19, 49-15

Classification

    Heron's Formula: 51-19
    Similar: 48-15
Chords: 46-28, 47-30
Circle:
  Inscribed Angle: 46-17
  Tangent to: 47-11, 47-18
  Tangents from a Point: 47-21, 51-17, 51-24
  Tangents, External: 46-17
Circular Arc Length: A72
Circular Arcs: 46-17
Circumcircle of a Triangle: 48-26, 50-21
Conic Section:
  Parabola: 49-14
Counting: 46-9, 51-25
Cube: 46-6, 46-30 (A95), 47-10
Cyclic Quadrilateral: 48-26, 50-16, A71
Distance Formula: 48-3, 48-9, 1-16, A63
Hexagon: 47-19, 47-30, 50-7, 50-23, 0-25
Inradius of a Triangle: A53
Lattice Points: 1-16
Line: 47-18, 0-8
Midpoint of Segment: 47-11, 47-17, 48-25, 50-3, 0-1
Parallelogram: A88
Perimeter:
  of Polygon: 48-2, 50-26, 0-11
  of Rectangle: 48-5
  of Triangle: 1-5, 1-7
Plane: 46-30 (A95), 47-28
Polygon: 46-17, 47-21, 48-2, 50-26
Power of a Point: 51-24
Pythagorean Theorem: 46-8, 46-23, 46-28, 47-9, 47-11, 47-20, 47-23,
    47-28, 49-28, 50-19, 50-21, 0-22, A56
Quadrilateral:
  Angles of: 49-26, 0-4
  Circumscribed Circle of: 48-26
Rectangle: 46-21, 47-15, 0-8
Sphere: 46-7, 47-27 (A96), 48-3

Solid Geometry: 46-6, 46-30 (A95), 47-9, 47-10, 47-23, 47-27 (A96),
   47-28, 48-23, 48-25, 49-27, 50-29, 51-10, A82
Surface Area of a Cube: 47-23, 49-27, A82
Transformational Geometry: 51-10
Trapezoid: 47-30, A59
Triangle:
   30-60-90: 46-18, 46-19, 47-17, 47-19, 48-19, 49-26, 1-7
   Circumscribed Circle: A90
   Equilateral: 46-19, 50-2, A72
   Inequality: 1-10
   Inscribed Circle: A53
   Isosceles: 46-17, 47-21, 50-19, A90
   Isosceles Right: 46-9
   Obtuse: 46-23
   Right: 51-21
   Similar Right: 46-8, 46-26
Volume:
   of Cone: 49-18
   of Cube: 0-9
   of Polyhedron: 48-23
   of Right Circular Cylinder: 49-18
   of Sphere: 49-18
   of Tetrahedron: 50-29

Number theory

Alternate Representations: 50-25
Base 10 Integer: 46-11, 46-13, 47-1, 48-13, 48-29 (A97), 49-3, 50-6,
   50-11, 51-23, A52, A58, A66
Base 2 Integer: 48-30
Cryptogram: 0-18
Digits of Integers: 46-11, 46-13, 47-1, 48-29 (A97), 50-6, 0-12, A52,
   A58, A73
Diophantine Equation, Nonlinear: 48-28, 50-30, 51-13, 51-16, A73
Divisibility: 46-11, 51-9 (1-14)
Divisors, Number of: 47-29
Factoring: 49-6, 49-13, 51-12
Fibonacci-like Sequence: 51-4 (1-6), A94

Geometry: 0-23
Linear Diophantine Equations: 48-22, 49-3
Modular Equivalence: 46-27, 48-20, 50-4, 0-14, 51-18, 1-17
Parity of Integers: 46-15, 46-20, 47-12, 48-10, 0-14, 0-15, A60, A91
Pascal-like arrays: 46-11
Place Value: 46-11, 48-13, A52, A58, A66
Prime: 0-7, 50-4, 51-6 (1-11)
Prime Factors: 48-28, 49-6, 49-12, 49-13, 51-1 (1-1), 51-7,
Pythagorean Triples: 46-22
Sum of Digits of Integer: 47-11, 47-14, 0-6
Unique factorization (FTA): 46-24, 46-29, 49-30

Statistics

Arithmetic Mean: 46-1, 46-25, 48-6, 48-7, 48-11, 48-18, 49-9, 50-20, 0-3, 51-14 (1-23)
Median: 46-25, 47-4, 48-18, 0-3, 51-14 (1-23)
Range: 46-25

Trigonometry

Cosine function: 49-19, 49-26, 50-18 (A99), A81
Definitions of Trig Functions: 51-17
Identities: 50-15, 50-27
Law of Cosines: 49-26
Law of Sines: 46-18, 49-26, 49-28, 51-17
Sine function: 48-27, 49-19
Trigonometric Equation: 50-15, 50-27

About the Editor

Harold Reiter received the PhD in mathematics from Clemson University. He has taught at the University of Hawaii, University of Maryland, Kingston University (London, England), and Clemson University. He has taught at the University of North Carolina Charlotte since 1972. He is the winner of several teaching and service awards including the Southeastern Section of the American Mathematical Association Distinguished University Teaching Award and the Paul Erdös Award of the World Federation of National Mathematics Competitions. He has served on the MATHCOUNTS Question Writing Committee, the Educational Testing Service's SAT II Mathematics Committee, and the CLEP Pre-calculus Committee. Together with his daughter Ashley Ahlin, he edits the problem section of the *Pi Mu Epsilon Journal*. He enjoys racquet sports, bridge, travel, reading and composing and solving mathematical problems. His main research interest is combinatorial games.